Using Museums to Popularise Science and Technology

Edited by

Sharyn Errington (Editor)
Susan M. Stocklmayer (Assoc. Editor)
Brenton Honeyman (Assoc. Editor)

COMMONWEALTH SECRETARIAT

Commonwealth Secretariat
Marlborough House
Pall Mall, London SW1Y 5HX, United Kingdom

Published by the Commonwealth Secretariat
Designed by K C Gan
Printed in the United Kingdom by Formara Limited
Wherever possible, the Commonwealth Secretariat uses paper sourced from sustainable forests or from sources that minimise a destructive impact on the environment.

Price £9.95 ISBN: 0-85092-668-8

Web sites: http//www.thecommonwealth.org
http//www.youngcommonwealth.org

Contents

Preface

A large number of Commonwealth countries are making efforts to popularise science and technology to achieve scientific and technological literacy amongst the public at large. A number of strategies are being employed to reach different target groups. In order to assist member countries in their efforts, the Commonwealth Secretariat has started a project on popularisation of the culture of science and technology.

Regional Expert Group Meetings were organised to identify the efforts which have already been made by different countries, problem being faced by them, ways forward and strategies that could be employed in the popularisation of science and technology programmes. The Regional Expert Group Meetings made a number of recommendations, which are being followed by the Commonwealth Secretariat and individual countries. A number of case studies from Asia have been published and a similar set of case studies is being developed for Africa.

The present booklet is a further contribution to existing Commonwealth countries in their efforts to popularise science and technology. One of the significant features of this booklet is that it highlights that not only science museums can be used to popularise science and, but other types of museum also have important contributions to make.

On behalf the Commonwealth Secretariat, I wish to express gratitude to the editor, Ms Sharyn Errington and Associate Editors, Ms Susan Stocklmayer and Mr Brenton Honeyman for putting this publication together, and to my colleague Dr Ved Goel, Chief Programme Officer, Science Technology and Mathematics Education for planning and managing it.

We hope that this booklet will assist museum curators and science teachers to utilise the full potential of museums in popularising science and technology.

Professor Stephen Matlin
Director
Human Resource Development Division
Commonwealth Secretariat

Foreword

The cane and the blackboard, the maths formulae and the exam schedule – all have long been part and parcel of the formal education system all over the world. Replace the cane with fun, the blackboard with computer multi-media, the maths formulae with hands-on exhibits, the exam schedule with a programme schedule – and you have a science centre! Teaching-learning processes in a science centre develop mental imagery through perceptual experience, and cognition through logical thought processes. Hands-on exhibits lead to minds-on experiences. In a science centre, imagination soars high and 'learn-through-fun' becomes the motto.

This publication is a global reflection of efforts made to utilise science centres, museums and other cultural institutions as a rich resource for public science and technology education. Exhibits and activities, new information technologies (both off-line and on-line) attractive presentation – all of these help in enhancing the creative faculties of the mind. No doubt, the learning process is informal but it supplements the formal curricular programme in such a way to break down the artificial barrier between informal and formal education.

The usefulness of this publication will depend on how much it will stimulate curators and programme designers in centres and museums to create experiences relevant to formal school programmes, and how much it will stimulate teachers to make use of the experiences available in these institutions. I am confident that this publication will do both.

Saroj Ghose
Past President, International Council of Museums, and
Retired Director General, National Council of Science Museums of India.

Introduction

This book is an initiative of The Commonwealth Secretariat, published as a response to the increasing popularity of informal centres of learning. Museums and centres of all kinds – including science, arts and culture – play a key role in educational infrastructure through the provision of programmes and experiences that facilitate the learning of science and technology in formal or informal contexts. At the same time, however, there is a growing awareness that visits, especially by school groups, could be more successful if there were greater interaction between museum curators and teachers. The Commonwealth Secretariat has produced this book to share ideas and approaches, not only for curators and educators working in museums and centres but for teachers in schools, in order to exploit more fully the potential of museums and centres in popularising science and technology.

A recent report by the House of Lords (House of Lords, 2000, ***Third Report of the Select Committee on Science and Society***) states that the future wealth and welfare of society depends critically upon the enthusiasm of young people to pursue scientific careers. "Science, technology and engineering are intimately linked with progress across the whole range of human endeavour: educational, intellectual, medical, environmental, social, economic and cultural... Science and engineering also make a most important contribution to improved public services and the quality of life."

None would disagree that young people should be encouraged to take up science, even if they do not intend to make it their career. The Report continues, "democratic citizenship in a modern society depends, among other things, on the ability of citizens to comprehend, criticise and use scientific ideas and claims… It is important that children do not grow up frightened of science and technology. Therefore a vital point of collaboration between the public, the media and the scientific community must lie in primary and secondary schools."

The UNESCO World Conference on Science in Budapest in July, 1999 made a similar call for a closer relationship between schools and the world of science. "Governments should accord highest priority to the improvement of science education at all levels" says the Framework for Action, "with particular attention to the elimination of the gender bias and bias against disadvantaged groups, raising public awareness of science and fostering its popularisation."

Science centres and museums have been presenting science in a different, informal format for many years but, too often, the school visit is not fruitful. For effective learning to occur – and within this book there are debates about the meaning of "effective learning" – there needs to be successful interaction between teachers and the museum, to facilitate and improve the visit of the school group.

The articles in this book have been provided by a variety of authors from many different centres. They come from many different countries of the Commonwealth where they have assessed local needs and addressed them through a variety of outreach programmes which are described in Section 1. Ways in which teachers can increase the effectiveness of a visit are outlined in Section 2 and formats for evaluating that visit are discussed.

The question of new technologies is a vexing one for science centres and museums. Such technologies are expensive, and few centres can afford to renew high-tech exhibits frequently. Yet to ignore their potential is to lose an important and popular dimension of the science centre experience. These issues are outlined in Section 3.

Research into effective learning in science centres and museums is increasing and there are some research groups for whom this is a major interest and concern. Three such groups are represented in Section 4, while others are scattered through the book.

The team who have edited this book are collaborators in a fruitful partnership between the National Centre for Public Awareness of Science (CPAS) and Questacon, the National Science and Technology Centre in Canberra, Australia. Questacon was the first such centre in the southern hemisphere and CPAS was built upon a long-standing successful and unique graduate programme in science communication at the Australian National University. CPAS is now also a Centre for the Australian National Commission for UNESCO. The long association between Questacon and CPAS through our joint graduate programmes and international outreach has led to our use of the term "public awareness" instead of "public understanding". We believe that this allows for a variety of approaches to issues at the interface between science and the public and allows for patrons of a science centre to enjoy the experience without the necessary condition of a formal learning outcome.

Teachers have a major role in changing the views of children – but the young students who are the main focus of this book go home each night to parents who also have a view about science and technology and whose visits to the science centre or museum are also to be encouraged. Only through such multi-faceted approaches will attitudes begin to change. Science centres have a unique opportunity to capitalise on those changing attitudes, encouraging families to follow up on school experiences. If that is to happen, however, the school visit must be enjoyable and engaging. The object must be to excite and enthuse. We believe that the activities described in this book fulfil that objective and provide a blueprint for teachers and curators to make the very most of visits by their young clients.

The Editors would like to thank Geoff Crane, Simon Yates and Julie Foster for their contribution in preparing material for this publication.

You can find Questacon and CPAS at the following websites:
www.questacon.edu.au
www.cpas.anu.edu.au/

Dr Sharyn Errington (Editor)
National Centre for the Public Awareness of Science
Faculties Teaching Centre, Australian National University,
Canberra, ACT 0200, Australia
Email: sharyn.errington@anu.edu.au

Dr Susan M. Stocklmayer (Associate Editor)
Director, National Centre for the Public Awareness of Science
Faculties Teaching Centre, Australian National University,
Canberra, ACT 0200, Australia
Email: sue.stocklmayer@anu.edu.au

Mr Brenton Honeyman (Associate Editor)
Manager, Public Programs and Education
Questacon – The National Science and Technology Centre
King Edward Terrace, Canberra, ACT 2600, Australia
Email: brenton.honeyman@questacon.edu.au

Section 1

Interesting Initiatives – Stories of Success

This section of the book is a selection of successful activities which have enhanced the visitor experience in a variety of settings. The authors are listed alphabetically, forming a random selection for the reader to sample.

In the first article, *Sharifah Barlian Aidid* outlines how the Malaysian National Science Centre decided to publish books for children, which the staff use in an interactive way with parents and children. Other in-house initiatives in science museums are described by *Paul Donahue and Claude Faubert*, whose Fun Days and Curriculum Days at the Canada Science and Technology Museum serve audiences of schoolchildren in very different ways. Bringing school students into the museum environment is also the topic of *Shannon Gilberg's* article, in which she describes summer camps and a popsicle building contest which have proved very successful outreach activities for the Fraser-Fort George Regional Museum.

Ways of bringing the natural environment closer to the science museums are discussed by *Chantal Barriault*, of Science North, and *Monty Reid*, of the Canadian Museum of Nature. Chantal's "Nature Exchange" has stimulated visitors and provided rich learning opportunities for adults and children alike. Faced with a different problem, Monty Reid describes how the Museum has embarked upon an ambitious project to collect data, research and educate about the biodiversity of the Rideau River.

Two articles address the problem of attracting new audiences to a botanic garden environment. *Julie Foster* describes how the Australian National Botanic Gardens set up tracks linking plant and animal life in ecological plantings, with good outcomes for younger visitors. *Janelle Hatherly*, of the Royal Botanic Gardens, Sydney, describes using the newly discovered Wollemi Pine to mount a thematic display and how the Gardens made the most of increasing public interest in indigenous culture.

Major outreach programmes require special support and funding. Two such programmes are described, involving many different organisations. The Caribbean Youth Science Forum was an initiative of the National Science Centre, Trinidad and Tobago. *Althea Maund* discusses the rationale for the Forum and describes its structure. From the opposite side of the world, *Allen Rooney* outlines a major travelling programme of Questacon, the National Science and Technology Centre, which reached many Pacific island nations.

Finally, four authors address wider issues of public activity. *Karen Fifield* of Taronga Zoo gives an overview of the possibilities of zoo education. *Geoff Snowdon* from PETROSAINS Discovery Centre presents a vision for a science centre. *Susie Bioletti and Philippa Winn* present a perspective on art and science through the lens of the National Gallery of Australia.

The overwhelming conclusion from these articles is that there are many approaches to building bridges with the public which have the potential to be successful. Those described in this section offer a model for others to follow, to adapt and to change to suit many audiences in many different countries. All have responded to local needs and are excellent examples of outreach which really works.

The Malaysian National Science Centre and its Science Stories as a Visitors' Companion Programme

Sharifah Barlian Aidid
National Science Centre, Bukit Kiara, Malaysia

Overview

The National Science Centre of Malaysia is situated at a 20-acre site at Bukit Kiara, Kuala Lumpur, Malaysia. It is an informal education institution under the Ministry of Science, Technology and the Environment. The Centre's mission is to inculcate a science-cultured society, which will contribute to the nation's development to improve the quality of life.

The National Science Centre of Malaysia

To implement the said mission, the Centre has a five-year strategic plan. The strategies are:

- to provide an environment as well as facilities for the fun teaching-learning of science
- to provide and build interactive science exhibits
- to provide and run interesting science programmes and activities
- to author and publish science literature
- to facilitate and advise on informal science education.

The Centre is equipped with various world-class facilities such as science laboratories, library, theatres, mechanical and electronics workshops, computer teaching laboratories, auditorium and seminar rooms. It provides an exciting informative atmosphere for visitors to explore, discover and learn the relationship between science, technology and everyday life without forgetting that a balance should be maintained between science, the environment and human ethics.

A total of 3,000 interactive, interpretive exhibits are housed in 14 thematic galleries, including a tropical freshwater aquarium and an outdoor educational science park. The design of some of the exhibits takes into consideration the limitations of the physically challenged visitors to the Centre.

The Centre carries out value-added activities under its *Science Enrichment Programme* known as *Program Tarbiyah Sains*. Both in-reach and out-reach activities are implemented throughout the year under this programme. Through the Centre's mobile science exhibition, the Malaysian rural communities get a taste of a science museum. Activities for pre-school children are designed in accordance with their cognitive development and space for these activities was designed with young visitors in mind. The teaching-learning techniques used in most of the activities are student-centred, informal and problem-solving.

The Centre authored and published its own bulletin, *Sahabat Sains*, (*Science Pal*) which parents and teachers can use to assess the effect of science centre visits on their children or students. The Centre has recently embarked on the writing and publishing of a series of science story books in the Malaysian language, which is the only one of its kind in Malaysia. This series of storybooks has been used by the Centre as its visitors' companion programme which is described in detail below.

Since the establishment of the Centre ten years ago, we have worked closely with university academics, educators and schoolteachers. The Centre conducts training programmes for trainee teachers, and provides research facilities to undergraduate and postgraduate university students. This research is focused on the effective mode of imparting and presenting science to the young and the general public. The Centre also provides an industrial training programme to undergraduate university students and credit points collected are accredited towards their degree.

Visitors' Companion Programme

The Centre has recently introduced a new visitors' companion programme, *Story Telling with the Centre*, after a great response was received from the parents and teachers from the *Science Visitors' Companion Programme*. There is great shortage of storybooks in the Malaysian language and no science storybooks at all, except those translated from other languages. Because of this, the Centre decided to write and publish its own series of science storybooks, in the Malaysian language, based on the science concepts as presented in some of the exhibits at the Centre.

Three science concepts – animal senses, animal camouflage and animal inter-dependence – were chosen. The moral of the stories is to inculcate love and respect in humans towards nature. The books are targeted for children between six and ten years, with the aim to cultivate the parent-child relationship through reading the books.

It was decided that the writers should be the Centre staff so they would have the opportunity to improve their science knowledge, writing skills and creative presentation of science concepts. This is also one way to help the staff to be multi-skilled, versatile, self-confident and develop self-worth.

A series of workshops on creative writing and graphic animations was organised by the Centre for the would-be writers and graphic designers. Professionals, academics and consultants were engaged to guide and to train the staff. After the training session the team was given a free-hand to develop their own ideas, story lines, style and format. A team spirit was maintained throughout the year it took to complete the project, which resulted in three books entitled:

- Bagaiman Ikan Mengesan Benda (How Fish Detect Things)
- Saling Bergantungan (Interdependence)
- Haiwan Menyamar (Animal Camouflage)

The Centre then co-published the three books with Dewan Behasa dan Pustaka, a renowned publisher in Malaysia. The books were officially launched by the Deputy Minister of Science, Technology and the Environment.

As an added value for the Centre's visitors and the readers, the stories from the books have been dramatised by the Centre staff for its *Story Telling with the Centre* programme. Through this programme parents learn how to use the books interactively and dynamically so that the science concepts are transmitted across to the readers through fun play.

Achievements

Since the Centre opened its door three and a half years ago, it has reached out to no less than one million people in Malaysia. The Centre hopes to contribute positively to the Nation's mission of generating a society of educated people.

Sharifah Barlain Aidid *is the Director of the National Science Centre Ministry of Science, Technology and the Environment, Pesiaran Bukit Kiara, Bukit Kiara 50662, Kuala Lumpur, Malaysia.*

Email: sharifah@psn.sains.my

The Learning Experience in *The Nature Exchange*

Chantal Barriault
Science North, Sudbury, Ontario, Canada

Science North is a world-class science centre located in Northern Ontario, Canada. One of Science North's most popular exhibit spaces is the trading area called *The Nature Exchange*. This is a unique learning environment that relies on the personal "trading" of *found* natural items and information about them, to motivate interest and understanding about the natural world. It has continuously evolved since 1984 and, during that time, compelling evidence indicates that this experience is fun and richly rewarding. It effectively raises awareness of key issues in the natural world around us and, through personal interaction, changes attitudes and behaviour. How does it work?

Bring it in and share

Visitors bring their natural collected items into *The Nature Exchange*. Trained staff initiate a discussion with the trader: what the object is, where it came from, why it is special and any other pertinent details. Traders are also encouraged to further their knowledge of their natural items through research in an adjacent resource library.

Children share the delightful findings in *The Nature Exchange*.

The Exchange

"Points" are awarded by the staff member based upon a series of criteria. Some of these criteria are based on the ethics of collecting and regional availability, while others are negotiable based upon each trader's enthusiasm, initiative and determination, as well as the following guidelines:

Information: What knowledge can you share with us about your item?
Uniqueness: What distinguishes your item from other similar items?
Quality: What condition is the item in?

Traders can have a direct impact on the value of their exchange. Even the simplest object can be awarded a high point value, if the trader shows initiative. The process of trading provides the necessary motivation. In searching for things to collect and trade, individuals learn to observe, to ask questions, and to think independently.

The Account

"Points" can be used immediately or banked in The Nature Exchange computer trading data bank and used for future trading for items in the Exchange's constantly rotating "collection" of goods.

Some artefacts need a closer look.

Take It Home

Visitors take their traded items home with them to add to their permanent collections, or bring them back in to trade for other items at a later date.

The learning opportunities provided by *The Nature Exchange* are abundant. According to Alan Nursall, Director of Science Program, Science North, "The best thing about *The Nature Exchange* is that most of the real learning takes place outside of the centre... where our visitors actually live their lives. Kids, and adults, too, get involved in actively observing, collecting, and studying the world around them. Science North, through *The Nature Exchange*, provides the venue that brings them together. It provides context and feedback, and is an outstanding tool for encouraging scientific behaviour in everyday life." Anecdotal evidence of the learning experience in *The Nature Exchange* is supported by a recent study, which examined the science centre learning experience more closely.

Learning has been described as "changing through experience....acquiring relatively permanent change in understanding, attitude, knowledge, information, ability and skill through experience" (Wittrock, 1977). In order to better understand the nature of learning and begin to recognise it when it is happening, a recent study was undertaken at both Techniquest, in the UK, and Science North in Canada (Barriault, 1998; 1999). This study investigated the behaviour of visitors as they interacted with exhibits to determine if there were consistent patterns of behaviours that occur which indicate learning is taking place. The investigations were carried out through detailed observation and interviews.

What was discovered is that there seems to be eight discrete learning behaviours that occur as part of the interaction with exhibits, and that these behaviours can be grouped into three categories that reflect increased interaction and depth of involvement.

Initiation behaviours:

- Testing out the activity
- Spending time watching others engaging in the activity
- Information and assistance offered by staff or other visitors

Above all else, visitors need to "feel safe" about committing themselves to engagement with an activity, especially in a public setting. Initiation behaviours enable them to "test the waters" with minimum personal risk and can be seen as the first step in learning.

Transition behaviours:

- Repeating the activity
- Expressing positive emotional responses in reaction to engaging in the activity

Smiles and outbursts of enjoyment along with repetition indicate that a level of comfort has been achieved and that visitors are comfortable, and even eager, to engage themselves more thoroughly in the activity. Regardless of whether the activity is repeated in order to better understand it, to master the functions or to observe different outcomes, the net outcome is a more committed and motivated learning behaviour.

Breakthrough behaviours:

- Referring to past experiences while engaging in the activity
- Seeking and sharing information
- Engaged and involved: testing variables, making comparisons and using information gained from the activity

Each of these behaviours acknowledges the relevance of the activity, and the learning gained from the activity, to the individual's everyday life. A personal level of comfort has been established that encourages a free flow of ideas and exchanges, and enables real learning to occur.

The Learning Behaviours in *The Nature Exchange*

An analysis of the visitor experience in *The Nature Exchange* indicates that visitors rapidly pass through the first two behaviour categories and that high levels of "Breakthrough Behaviours" occur on a regular basis.

- Visitors involved in trading commonly refer to past experiences when presenting an item for trade, either by mentioning where it had been found or how they discovered its identity. For example, children describing their items to the staff person explain that: "I found a beaver stick when we were camping in French River..." or "I found this in a raspberry patch near my house".
- Seeking and sharing information are behaviours that permeate the entire experience, both before and during the visit. Visitors seek information about their trading items before coming to The Nature Exchange and are eager to share this information with the staff person. Information shared ranges from "this rock has a lot of colours" to "this is fluorite...when we scratch it with a metal nail, the hardness test says it's fluorite".
- Engaged and involved is exemplified most profoundly when visitors describe their items, discuss the observations of their item and draw conclusions about their item with the staff person. For example, a young boy pointed out beaver teeth marks on the stick he brought in, while another visitor, trading clam shells explains, "I know that starfish eat them and sea gulls too. They open them and eat them". The trading activity is very involved, lasting from five to 40 minutes. This level of engagement and motivation indicates a rich learning experience about the natural items brought in by the traders and encourages a commitment to learning outside the science centre and into the home environment.

The success of *The Nature Exchange* as a motivating learning environment has encouraged other centres to open a Nature Exchange of their own. *The Nature Exchange* is being offered by Science North as a turnkey package to other institutions. The Dallas Children's Zoo has adopted a Nature Exchange with great enthusiasm and success. For more information about *The Nature Exchange*, visit our web site at www.natureexchange.com.

References

Barriault, C. (1998). The Science Centre Learning Experience: A Visitor-Based Framework. MSc Thesis, University of Glamorgan / Techniquest, Wales, UK.

Barriault, C. (1999). The Science Centre Learning Experience: A Visitor-Based Framework. The Informal Science Review, 35, Mar-April.

Wittrock, M. (Ed.). (1977). Learning and instruction. Berkeley, CA: McCutchan

Chantal Barriault *is the Staff Scientist – Education Initiatives at Science North 100 Ramsey Lake Road, Sudbury, Ontario P3E 5S9, Canada*

Email: barriault@sciencenorth.on.ca

Art and Science? ... NGA!

Susie Bioletti and Philippa Winn
National Gallery of Australia, Canberra, Australian Capital Territory, Australia

As the premier art institution in Australia with over 90,000 works of art in its collection, the National Gallery of Australia (NGA) recognises that along with preservation and display, education and learning are integral to its purpose. People learn best when learning is fun and new information builds on what is already known and accepted. Science is fundamental to all aspects of life and provides a platform to build an understanding and appreciation of the visual arts.

A workshop at NGA about the science and techniques of conservation.

Tools and materials of the art of illumination on display during the exhibition of the Book of Kells.

Pigments and materials used in the art of illumination.

The National Gallery of Australia contributes to the annual National Science Week with a programme of popular events such as *Conservators on the Floor* and *Expose Yourself to Science*. These interactive demonstrations attract and inform primary, high school and tertiary students as well as a general audience. Conservation tools, such as microscopes and the infra red vidicon, demonstrate how conservators examine and assess the condition of works of art. Visitors are introduced to issues of preservation and storage with works from the national collection undergoing conservation.

The National Gallery Education Department provides art discussion tours tailored to the interests of the group. Tours, focusing on science and art, are designed and promoted for National Science Week and are available throughout the year. Popular requests for discussion tours include *Art and the Environment* or *Recycling and Art*.

Conservators at the National Gallery of Australia regularly contribute articles to *artonview*, the National Gallery of Australia's quarterly magazine. These articles focus on materials and techniques used to create works of art. Written for the general reader they include the results of scientific analyses and anecdotal information. Conservators also present public lectures on a regular basis. These contributions and lectures engage the reader in the physical aspects of art and alert them to the key conservation projects undertaken in the laboratory at the Gallery.

Prominent scientists, science writers and broadcasters are invited to be part of the NGA Public Program. For National Science Week 2000 Australian Broadcasting Corporation (ABC) personality Dr Norman Swan spoke about the meeting between science and art.

In Canberra art and science collaborate formally through "Metis" the exhibitions of science and art programme instigated in 1999. The National Gallery participated in "Metis" with an exhibition of Paula Dawson's holographic work *There's no place like home*. In 2000, "Metis II" will take the theme of waste and the National Gallery of Australia will again participate in this dynamic programme.

The recent exhibition, *The Book of Kells and The Art of Illumination*, incorporated pigments and tools to draw attention to the technical aspects of creating illuminated manuscripts. The science–art components of exhibitions and events at the Gallery, such as garden design and botany discussion tours of the Sculpture Garden and the Fiona Hall Fern Garden are always well attended and have proved extremely popular.

Susie Bioletti *is the Senior Paper Conservator*
Philippa Winn *is Project Officer Education and Public Programs*
The National Gallery of Australia, King Edward Terrace, Canberra, ACT, 2600, Australia

Email: philippa.winn@nga.gov.au

Making Heritage Relevant

Paul F. Donahue and Claude Faubert

Canada Science and Technology Museum, Ottawa, Ontario, Canada

The Canada Science and Technology Museum Corporation is comprised of the Canada Science and Technology Museum, the Canada Agriculture Museum and the Canada Aviation Museum. It is mandated "to foster scientific and technological literacy throughout Canada by establishing, maintaining and developing a collection of scientific and technological objects...and by demonstrating the products and processes of science and technology and their economic, social and cultural relationships with society." The corporate mission statement is: "To discover and share knowledge about Canada's scientific and technological heritage in order to increase understanding and appreciation of the role that science and technology has played and continues to play in the transformation of Canada." This paper focuses on the educational programmes of the Canada Science and Technology Museum, the only general science and technology museum in Canada.

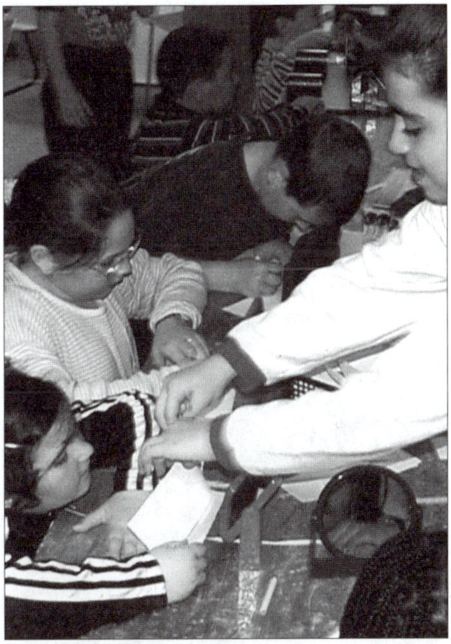

Fun with mirrors!

Locomotives from times past are on display at the Canada Science and Technology Museum.

It should be noted at the outset that education is a provincial responsibility in Canada. Formal science education is generally focused on the present, with the goal of producing more scientists, engineers and technologists, and the corporate mandate is to preserve and interpret Canada's scientific and technological heritage. The challenge is to meet the needs of students and teachers while fulfilling the museum's mandate and mission.

In 1999 the museum offered guided programmes to over 50,000 schoolchildren. More than 60,000 other students came to the museum for a general visit. The number of students who participate in programmes has been growing over the past few years and the increase can be attributed to the development of innovative programmes that meet the needs of the teachers while preserving the unique character of the museum. Teachers who bring their students to the museum do not want a classroom experience for them. At the same time, they have a curriculum to cover and also face a lack of resources and great demands on everyone's time. Teachers, therefore, are looking for experiences that offer great relevance to the science and technology material that must be covered in class and that are presented in fun and stimulating ways.

The museum is fortunate in that it presents real objects from Canada's rich scientific and technological past to the public. It displays locomotives and cars, computers and bicycles and also offers students controlled access to the real thing. This "real" environment can only enrich the students' visit. Furthermore, the programmes give the students ample opportunity to participate and experiment. Because of its dedicated resources and experienced and knowledgeable staff, the museum offers programmes that go beyond what many schools can offer.

Staff at the museum guide young visitors through interactive activities

The museum quickly embraced the internet (www.nmstc.ca) and has over the last two or three years, created a number of pages to support its school programmes. The pre-visit and post-visit kits that are available to teachers in print form are now also available on the museum's web site. The museum has also developed a number of pages that offer supplementary information on a range of topics of interest to teachers and students alike. The number of visits to these pages has increased regularly over the past few years and the feedback received from teachers seems to indicate that they are making great use of the information available.

Examples of some of the efforts undertaken in these innovative programmes are *Fun Days* and *Curriculum Days*.

Fun Days

All who work in a museum or science centre in North America are familiar with the invasion of students that happens every May and June as the school year comes to an end. Of late, schools have had their budgets cut dramatically. This has resulted in approval for school trips having become more difficult to obtain. Five or six years ago, the museum's manager of school programmes suggested that we offer programmes that combined learning and fun in a unique way. It was thought that teachers would be more than happy to put together a school trip to the museum at the end of the school year if their students could participate in activities related somehow to the science curriculum.

When Summer Fun Days were first introduced they were an immediate success. The format is quite simple: the museum offers between six and ten activities, each given simultaneously in a different part of the building. Each school group can select two of the 45-minute activities. All activities start and end at the same time. The groups then move on to their next selected activity. Each activity begins with a short demonstration performed by one of the museum educators. The rest of the activity is dedicated to hands-on experimentation. The content of each activity is loosely based on the Ontario grades 1 to 6 science and technology curriculum. For example, in one activity for grades 1 to 3, students use basic forensic skills such as fingerprinting, identification of footprints and microscopic examination of samples to help catch a museum thief.

In another activity, recommended for grade 4 to 6, students build their own space station while finding out about the International Space Station, the *Canadarm* and Canada's role in space exploration. This particular activity takes place in the Canada in Space hall, so students can see many of the objects mentioned by the educators. Holding the activity in the Museum's galleries not only offers an unusual teaching space, but it allows repeated contact with the Museum's rich collection of artifacts. In June 2000, the museum held eight days of Summer Fun Days and received over 4,000 students.

These programmes offer students the chance to do hands-on learning activities, through our equipment, environment and knowledgeable staff. The museum staff incorporate as much heritage content as is possible, including using the museum as the "classroom" and visiting some exhibitions as part of the activity itself. The result is fun, education and a lot of activity within the museum over the 90 minutes of participation. On those days the museum is not a place of quiet introspection, but rather one full of supercharged adolescent energy.

Curriculum Days

In 1998, the province of Ontario introduced a new science and technology curriculum. Most teachers were not adequately prepared to teach the new curriculum and resource materials were not available to them. At the time, one of our educators proposed that we offer a series of new programmes, designed to present to students specific aspects of the new curriculum. This series of new programmes followed more or less the same format as Fun Days and was offered to students from grades 1 to 6. The main difference is that the activities are grouped in pairs for each grade level. The teachers cannot pick and choose as in Fun Days. For example, grade 4 students take *Pulleys and Gears*, and *Light*.

In the programme *Light*, students face a new challenge a each of the stations and work co-operatively to solve a problem. By experimentation and exploration they see principles of light in action at the telescope, microscope, periscope and kaleidoscope stations. A visit to the Helen Sawyer Hogg Observatory wraps up this workshop.

In *Pulleys and Gears*, students explore gear systems to discover amongst other things the relationship between the number of teeth on a gear and its speed of rotation. Hands-on construction of gear systems illustrates the roles that gears can play in a system such as idler gear or spur gear. These concepts are further reinforced when the students view various artifacts in the museum's exhibits that illustrate gears at work. In March and April 2000, there were six Curriculum Days and over 3,000 students participated. The dates in March and April were chosen because they coincide, more or less, with the end of one science unit and the beginning of another one.

The response from teachers and students was overwhelming in favour of our efforts. Teachers who were not able to register for lack of room immediately register for the next year. Similarly those who did successfully register, pre-registered for the next year. If anything, the programme was too successful and in our relatively open museum of 10,500 square metres, the noise level became excessive and it was difficult to both hear and to concentrate. The following year we increased the number of days. We were able to register more students but there was less "bedlam".

Other initiatives

The museum offers as part of its regular programmes a number of demonstrations that highlight its collection. These demonstrations are open to all students, whether they are here on a general visit or to take part in a school programme. For example, volunteers from the Bytown Railway Society operate a restored lumber yard steam locomotive every Wednesday and Sunday between May and October. Many schoolchildren are able to take a short ride on the cars pulled by the locomotive while staff guides explain to them how steam locomotives work and how this Shay locomotive came to be part of the museum's collection.

All year round, staff ride replicas of old bicycles around the museum. Schoolchildren often ask questions and many get to take a short ride on one of the replicas.

Finally, the museum's floor staff, called guides, constantly interact with schoolchildren when they are in the "free" part of their visit to the museum.

This blend of displays of outstanding real objects illustrating Canada's rich scientific and technological past, interactive exhibits, live demonstrations and classroom-type activities produces a rich experience for every one of the more than 110,000 students who visit us every year.

Dr. Paul F. Donahue *is Director General of Public Programmes for the Canada Science and Technology Museum Corporation, P.O. Box 9724, Station "T", Ottawa, Ontario, K1G 5A3, Canada*

Email: pdonahue@nmstc.ca

Claude Faubert *is Site Director and Director, Education and Interpretation, for the Canada Science and Technology Museum.*

Wild Ways of Learning: Zoo Education in New South Wales

Karen Fifield
Taronga Zoo, Sydney, New South Wales, Australia

Zoos are sensory places – children and adults can see, smell, hear and touch wildlife!

Zoos educate people across the whole of society, from all socio-economic and cultural backgrounds. They all do it in their own way even though there are some common themes.

For many people their connection with the environment happens in a zoo – especially in urban communities where experiences with nature are few and far between.

A group of primary school children interact with an Australian marsupial - the quokka.

Zoos are dynamic places – they have changed and are changing. Animal collections have been in existence since early times. However modern zoos are very different from their earlier counterparts. They are evolving into multi-purpose conservation centres.

Zoos are seen more and more as educational organisations – striving to get the message across about all this important work and to ultimately link people with nature and ignite the spark towards environmental literacy.

"Zoo Education has almost become an independent discipline. The use of a wide variety of educational techniques, facilities, and considerations, together with knowledge, creativity, and inventiveness can make zoos highly interesting, attractive, and effective places for environmental, conservation, and holistic life system education." (World Zoo Conservation Strategy, 1993)

Zoos worldwide are visited annually by over 600 million people which is more than 10% of the world's population (World Zoo Conservation Strategy, 1993) – visitation unequalled by any other group of public, conservation-oriented institutions. This is not a bad effort considering we do not have a

Tomorrowland, Frontierland, Adventureland or Fantasyland in sight! Zoos are ideally suited to deliver worthwhile education programmes because of such high visitation and interest from the community.

Educationally proactive organisations can influence public attitudes and highlight environmental issues in a variety of ways. The zoo community must help to reverse environmental trends through targeted education – organisations such as the International Zoo Educators Association, recognise the need for well placed education programmes, international co-operation and uniform messages about the unique role of our institutions.

Zoos exhibit living animals, which makes them different from museums and other cultural institutions – animals give zoos an edge. Zoos are in the perfect position to promote environmental education and through this to shape public opinion about conservation.

If we take Taronga and Western Plains as typical zoos we can examine the options available for educational experiences. Over one million people visit the two zoos of NSW each year. We aim to provide varying levels of educational experiences for our diverse range of visitors.

Informal Education – education not directly linked to the curricula of formal educational institutions

Public programmes and informal learning situations are important as contact points for the vast majority of zoo visitors. Informal education takes place through information labels, interpretives, brochures, media coverage, keeper presentations, school holiday programmes, playspaces, animal contact opportunities, touch tables, tours, special events, the animal themselves and environmental practices within the zoo.

Each of these informal situations must have the target audience clearly defined and messages made simple (even if the concept is complex) and easily understood. It is the role of educators and interpreters to make these informal experiences memorable – they may be the first key to increasing understanding and changing attitudes in our visitors.

Formal Education – more directly linked to curricula frameworks

Formal education takes place through publications, specialist lectures, adult education courses, the TAFE Zookeeping Course, other special interest group meetings (like the Veterinary Association) and the Australian Conservation Training Institute (for overseas conservation organisations) and school education.

These formal situations are targeted at specific audiences and deliver more complicated messages about the work of zoos. To give you an example:
Over 120,000 school students were involved in some way with Taronga and Western Plains Zoos last year – that is 10% of total visitation to the zoo. Of the 3,068 Schools in NSW in 1997, approximately 1,300 visited Taronga. Clearly a zoo visit is an integral part of the school curriculum. Environmental education is mandatory in NSW schools.

The draft 1999 NSW Department of Education Environmental Education document states that "environmental education is a lifelong process of learning that helps people to understand and appreciate the environment and their inter-connectedness with and responsibility for it." Zoos are critical to environmental education – "effective environmental education provides diverse learning experiences in and beyond the classroom".

Some of these experiences at Taronga Zoo are:

- Classrooms with live animals and interesting biofacts. Kids can touch, see, smell, hear and get close to animals. This goes a long way to breaking down fears and barriers towards animals. The lessons address the curriculum and are designed to complement the teacher's classroom programme.

- Special programmes which link the zoo to curriculum areas. The Writing Competition and Banner Competition require students to research an animal theme and develop a curriculum-based, zoo-linked outcome.

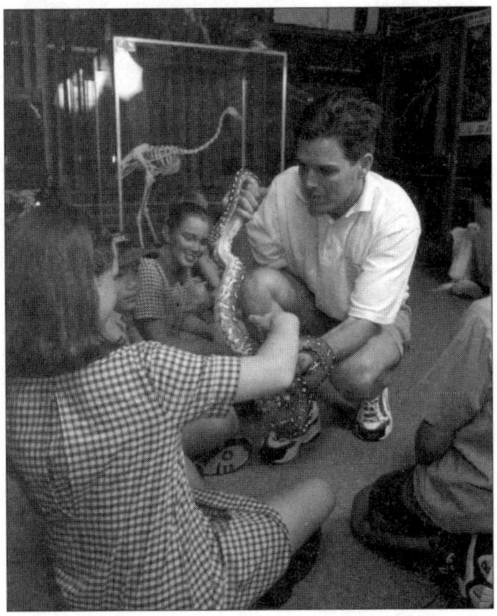

A close encounter with a live snake.

- The Zoomobile, our outreach programme (with our cool and 'out there' educator!), visits schools unable to visit the zoo (approximately 10,000 students each year). These include hospital schools, remand centres and isolated schools. Country trips to isolated schools cause a great stir as the animals check into the local motel!

- Animals of the Dreaming (another outreach option) combines interesting aspects of Australia, native animals and Dreaming stories which relate to a number of curriculum areas.

- Work experience programmes give students a taste of being a keeper and meet vocational education requirements for schools.

Searching for tadpoles in a creek.

- Zoo Adventures, our holiday programme, provides educationally based programmes with behind the scenes opportunities.
- ZoosnooZ – sleep at the zoo. What a wonderful way to get connected! Imagine getting close to Kotik, our 225 kg sealion, or feeding a giraffe! The teachers don't mind it either. It is a once in a lifetime experience.
- Research Week, where students research an area of the zoo, work with zoo staff and use the zoo's specialist library.
- Internet programs like Murder Under the Microscope, the zoo's website chat room.
- Wild Homes is a design and technology programme where Year 11 students design zoo exhibits with the assistance of the zoo's Exhibit Planning Officer and architects.
- Teacher inservice programmes to give teachers a better understanding of educational opportunities available at the zoo.
- Joint programmes with other environmental agencies like Streamwatch and State Forests.

But the real challenge for zoo educators has been and will be to take our expertise outside the zoo boundaries. As zoos develop more *in situ* conservation programmes it is imperative that community education programmes support this work.

The Taronga Zoo Education and Herpetofauna staff have developed a community based programme to reintroduce Green and Golden Bell Frogs to Sir Joseph Banks Reserve at Botany. We work with the local schools, council and community group to look after the site and collect scientific data about the tadpoles and frogs.

This programme has now been the catalyst to develop a national education programme for frogs (ASX Frog Focus) through the zoo education network – an Australian first. This educational interactive CD-ROM will give teachers and students across Australia information and activities which will help them help frogs!

Programmes like Frog Focus, which combines conservation and education, are at the cutting edge of future zoo education – local ownership of conservation is the most meaningful. Local communities are uncertain of where to begin even if they are motivated – it is here that zoo experts can help. By combining the talents of educators and life sciences staff, worthwhile conservation education programmes can develop.

This programme is zoo education at its best; after all, education is about opening doors to new experiences, new ways of thinking and new ways of behaving. What more is there to say except:

"In the end we will conserve only what we love, we will love only what we understand, we will understand only what we are taught". Baba Dioum

References

NSW Department of Education and Training (1999), Draft Environmental Education Policy.

World Zoo Conservation Strategy (1993) Chicago Zoological Society, Brookfield Illinois. IUDZB/CBSG.

Karen Fifield *is the Education Manager at the Toronga Zoo, P O Box 20, Mosman, NSW 2088, Australia.*

Email: kfifield@zoo.nsw.gov.au

Cultivating Green Awareness: Specialist Tracks and Programmes for Young Children at the Australian National Botanic Gardens

Julie Foster
Australian National Botanic Gardens, Canberra, Australian Capital Territory, Australia

Botanic gardens throughout the world are visited by over 150 million people each year and they provide an ideal environment for educating people on the beauty, importance and diversity of the plants in their lives.

The Australian National Botanic Gardens (ANBG) situated in Canberra, the capital of Australia, was opened in 1970. It occupies an area of 90 hectares close to the city centre and contains the world's largest collection of Australian native plants with more than 6,000 species in cultivation. Its objectives are to 'increase knowledge, appreciation and enjoyment of Australia's plant heritage'.

Face painting at the Australian National Botanic Gardens - become your favourite flower or animal.

About 380,000 people (including 40,000 students) visit the Gardens each year. Many visitors come in family groups which, include young children and it is for this group that the ANBG developed special self-guided tracks and activities based on one of the major display themes of the Gardens – the ecological plantings. These displays feature plants which grow together, such as plants in the Rainforest, the Sydney Basin and the Mallee Shrublands. Other themes include taxonomic displays, rare and threatened Australian plants and plants used by Aboriginal peoples.

Plants do not have the intrinsic interest to young children that animals have. However, botanic gardens can use this to their advantage by focusing on plants which provide food and habitat for animals. Ecological plantings provide the ideal opportunity for this.

Temporary Self-guided Children's Tracks

Leaftail – The Leaf-tail Gecko Track

Leaf Tail by Narelle Oliver (1989, McCulloch Publishing). The storyline from the book was the basis for a children's track in the Australian National Botanic Gardens.

This track followed the adventures of a Leaf-tail Gecko (lizard) in the Rainforest area of the Gardens. It was based on the children's book ***Leaftail, the Story of a Leaf-tail Gecko*** and was in place for three months including the summer holidays. This track was very successful. It used a simple story line and creative and original model animals with clear interpretive signs. The Leaf-tail Gecko models were so lifelike that a number were attacked and pulled to pieces by hungry kookaburras. At each stop along the track the story and some artwork from the book were displayed. The story could be read to younger children as they searched for Leaftail and friends – a frog, a pademelon or a bowerbird – up a tree or under a log. Activities related to the track were held during the summer holidays. The author of the book conducted workshops in creative writing and printmaking for children and adults.

What's for Lunch? Track - ACT primary school children looking for a koala.

What's for Lunch? Track

The What's for Lunch? Track was aimed at pre-literate children. Based on the popular children's book by David Miller, the track was uniquely suited to the Gardens' environment as it featured many local animals, which need the plants of the Gardens not only for food, but also for shelter and sometimes for nesting.

David Miller's paper sculptures, which are the basis for the vibrant illustrations in the book, were on display in the Gardens' Visitor Centre. A free activity book and map provided resource material for both children and adults. At intervals along the track the animals featured in the book could be found and the question asked – 'What do they eat for lunch?' The answer was under a flap on the displayed interpretive sign. David Miller conducted workshops showing children how to make their own paper sculptures during the summer holidays.

In the Visitor Centre children could also play with feltboards and felt animals and plants to make up their own pictures showing the inter-relationships of plants and animals. The feltboards had previously been used very successfully with major ecological exhibitions as a creative and educational diversion for children as accompanying adults viewed the exhibitions. A range of animal puppets from these habitats was also available for play.

Snugglepot and Cuddlepie

Another track for young children titled ***Snugglepot, Cuddlepie and their Friends*** meandered amongst plants which grow naturally on the sandstone soils of the Sydney region. The track used characters from a well-known Australian story book by May Gibbs and focused on the flowers, fruits, leaves and bark from a variety of plants of this region.

The Snugglepot and Cuddlepie Track - a scribbly gum letter.

Before commencing the track a story book, map and activity booklet were collected from the Visitor Centre. At each stop the accompanying adult read part of the story to the children and they made drawings or completed similar activities in their booklets. This type of track has a limited life as plants stop flowering or their fruit are eaten by parrots or possums, however coloured beads can be used to replace soft fruit such as Lilly Pilly and laminated pictures can be used in place of the flowers of boronias and wattles.

Temporary tracks can also be put in place to celebrate special events such as Wattle Week. The taxonomic wattle display is an excellent site for such a track.

The Wattle Week Track

About 600 of Australia's more than 1,000 species of Acacia are represented in the Wattle Section. The Wattle Week Track passes all kinds of wattles – those which have special flower or leaf features, those which were significant to Aboriginal people or early European settlers or those which have a significant place in Australia today e.g. Australia's floral emblem, Golden Wattle (***Acacia pycnantha*).** This track was designed for children who can read and the accompanying booklets contain a variety of activities.

The Wattle Walk - children feel the soft Wattle flowers.

Other programmes to engage children and help them learn the messages of conservation have included puppet shows, story telling and painting competitions. Another programme popular during holiday periods was Face Painting, in which children changed into flowers or animals. The children could also discover parts of the Gardens where the plants or animals lived using specially prepared activity sheets.

Setting up an interpreted track

It is important that discussions are held with all relevant staff before any decisions on a new interpreted track are made. This can help to ensure that potential difficulties can be dealt with early. Curators, gardeners and interpretive staff as well as those from visitor services, education and promotion could be involved. Paths may need upgrading or extra plantings required. Understandably gardeners and curators are often apprehensive about the idea of large groups of children moving through the areas for which they are responsible.

Prior to final installation of the new track and its promotion, it is important that people unfamiliar with the Gardens walk the track to test it. They can help to identify problems related to instructions, directions and interpretation. Tracks must be marked clearly and the accompanying maps need landmarks such as bridges or seats shown clearly to assist people with orientation.

Maintenance

Tracks must be checked daily and any signs cleaned and repaired or replaced if necessary. This is best done by staff who have installed the track. Some tracks can be very time consuming to maintain. Interference with and damage to directional signs and props are problems often encountered.

Evaluation

Many of the tracks prepared at the ANBG are in response to suggestions and requests from visitors. Visitors are encouraged to provide written feedback on their visit and to suggest any improvements to programmes, labelling and interpretation. An important part of the planning of an interpreted track includes a decision on the method and process by which it will be evaluated.

Julie Foster *was the Education Manager, Australian National Botanic Gardens Canberra, ACT. She is now the Manager, Visitor Programs, Questacon – The National Science and Technology Centre, King Edward Terrace, Canberra, ACT, 2600, Australia.*

Email: jfoster@questacon.edu.au

Broadening Science Perspectives

Janelle Hatherly
Royal Botanic Gardens Sydney, New South Wales, Australia

Botanic gardens play a significant role in raising community awareness of the importance of plants in our lives and the need for their conservation in the natural world. Spectacular living collections of plants, thematically arranged and well interpreted, provide general visitors and structured groups alike with a memorable immersion experience and a meaningful context for learning.

The Royal Botanic Gardens Sydney (RBGS) consists of three sites: the well-known gardens located in the heart of the city next to the Opera House on Sydney Harbour; a cool temperate display garden 105 kms to the west at Mount Tomah and an extensive collection of Australian native plants at Mount Annan Botanic Garden, 90 kms south west of Sydney.

RBGS has as its slogan, "Plants = Life", and delivers this contemporary message in relevant and focused research and publications, garden displays and public programmes. Each year over 3 million people visit the Gardens' three sites and over 18,000 students take part in interactive lessons. In addition, another 40,000 visitors participate in holiday activities, courses, talks and guided walks. Even more people access the Gardens messages remotely through the community education outreach programmes or, electronically, through the web site, which includes PlantNet, a comprehensive database of NSW flora.

In recent years RBGS has seized upon two unique opportunities to popularise science and technology and broaden people's perception of the work done by scientific institutions. The first came as a result of the botanical find of the century and the second because of rising public interest in indigenous culture and its links to the unique location of the Sydney Gardens.

The two case studies below illustrate how botanic gardens as living museums can 'make a difference' by providing provocative displays, programmes and experiences that facilitate learning of science and technology and different cultures.

Wollemi Pine – Dinosaur Tree

In late 1994, in a remote section of the Wollemi National Park just 150 kms from Sydney, a National Parks Ranger discovered one of the world's rarest plants. The uniquely Australian Wollemi Pine is only known by 38 adult plants

School children in Darwin (Northern Territory) with a two-year-old Wollemi Pine grown from seed.

in the wild and belongs to an evolutionary line of conifers thought to be long extinct. To ensure the Pine's continued survival the National Parks and Wildlife Service developed a conservation strategy for its natural habitat and RBGS undertook extensive horticultural research on mass propagation to make it available commercially by 2004.

While the Pine's rarity makes it inherently interesting, it is far more significant as a powerful stimulus, a 'wow' factor, for introducing concepts such as biodiversity, rare and threatened species, conservation, national parks and wilderness areas. It makes it easy to raise awareness of the importance of scientific research and what is achievable with the available technology. It also helps promote botanic gardens and museums as centres for biological research and environmental education.

In response to the great amount of public and scientific interest in the Wollemi Pine, the RBGS planted young Wollemi Pines in the Sydney, Mount Annan and Mount Tomah Gardens and made a commitment to mount educational displays. At the Sydney Gardens the Wollemi Pine is dramatically presented in a cage as part of a broader thematic display called *Plants in Peril* or *The Rare and Threatened Species Garden*. Supporting this display are regular educational activities and just two examples of these are on-site touch trolleys interpreted by volunteer Green Guides and Wolly's birthday parties (for 3-8 year olds).

The Community Education Unit also developed a travelling display and interactive slide presentation, *Wollemi Pine – A Dinosaur Tree*. It includes slides of the Pine in its natural habitat, a young living plant, fossils and specimens of bark, cones and leaves of mature trees which can all be handled. The presenters, all experienced education officers and horticulturists from the Gardens, adapt the programme to suit the ages and interests of the participants, and involve them in analysing the research findings and solving problems associated with protecting a wilderness site and propagating from a limited population.

Because it will be a few years before other institutions can have their own Pine specimens to interpret, *Wollemi Pine – A Dinosaur Tree* has toured extensively around Sydney, the State and interstate (the latter thanks to a grant from Visions of Australia, the Federal Government's touring exhibition programme) as part of the Community Education Unit's outreach programme. In this way, since 1996, approximately 10,000 people who could not visit the Gardens have been able to view and learn about this living treasure. The concept of 'a living dinosaur' has also attracted good local newspaper, radio and television coverage at each of the venues visited. This significant media interest has meant that many more people than those who saw the display have became aware of the Wollemi Pine, biodiversity and conservation issues.

Cadi Jam Ora – First Encounters

Cadi Jam Ora means "I am in Cadi". The land occupied by the Sydney Gardens holds special significance for Aboriginal people, as it is the site where the modern transformation of the Australian continent began. It can be regarded as the "first frontier" between Aboriginal and European societies. On this site in 1788, under the leadership of Governor Arthur Phillip, white settlers cleared the vegetated creeks and woodlands and set about creating a farm. These initial attempts to establish an agricultural foothold on Australian soil struggled and failed. Of equal significance, some of the earliest prolonged encounters with Indigenous Australians happened here, leading ultimately to the tragic demise of the local Cadigal people. By 1790 only three of the clan living around Sydney and Farm Cove were left.

Constructing a giant goanna on the Cadi Jam Ora site as part of a January holiday programme.

In 1998 the RBGS made a corporate commitment to Aboriginal reconciliation and identified specific ways of representing Aboriginal heritage and culture in displays and programmes. A four-bed garden display called *Cadi Jam Ora – First Encounters* was developed to explore Aboriginal people's prior use of the site and their understanding of plants and the environment. It interprets Aboriginal uses of plants for food, medicine, tools and cultural purposes and illustrates how the land the Gardens occupy was and is significant to Aborigines. While many cultural institutions around Sydney address this period in history, the RBGS is uniquely placed to interpret the role of specific plants in the local Sydney Aboriginal culture as well as the events that happened, on this very spot, at the time of white settlement/invasion.

As a scientific institution RBGS is also keen to acknowledge that different cultures hold different environmental and ecological perspectives. While the Eurocentric view of the world does not recognise as technology Aboriginal knowledge, skills, organisation, tools, artefacts and methodologies, Aboriginal approaches provided a more successful way of surviving in the Australian environment than the European approach adopted by the early settlers. *Cadi Jam Ora* and its associated activities compare and contrast indigenous technology with the European technologies used to establish the First Farm (which is now a part of the four-bed display).

The Gardens employed an Indigenous Education Officer to ensure ongoing consultation with local Aboriginal communities and to ensure *Cadi Jam Ora* conveys a contemporary indigenous perspective. Strong partnerships have been established and this has led to direct indigenous community involvement in many public programmes at all three RBGS sites.

RBGS hosts many events and exhibitions related to indigenous issues such as National Sorry Day and Journey of Healing, the Sea of Hands and the proposed City of Sydney indigenous sculpture walk. A living memorial to the Stolen Generations is being built at Mount Annan Botanic Garden, and Mount Tomah is starting an accredited ecotourism course for indigenous interpreters. The Community Education Unit regularly runs lessons for school students, holiday programmes and guided walks with an indigenous theme.

Interpreting the indigenous history of the Gardens' sites marks a significant new direction for the RBGS and is one of the first attempts by a botanic garden to place indigenous and modern cultures into the perspective of their environment.

Janelle Hatherly *is the Community Education Manager at the Royal Botanic Gardens, Mrs Macquarie's Road, Sydney, New South Wales, 2000, Australia.*

Email: janelle.hatherly@rbgsyd.nsw.gov.au

Caribbean Outreach – A Special Project

Althea Maund
National Science Centre, Trinidad and Tobago, West Indies

As a result of colonial heritage, the English-speaking Caribbean lacks a tradition and culture in science and technology. In the pre-independence era of most of the islands, science and technology were not considered critical elements of primary and secondary curricula. Since, at that time, our economic horizons did not extend beyond the production of primary products, education, as a whole, did not have as its major aim, the preparation of citizens with the skills and confidence to work towards the elimination of the client-dependant status.

Members of the audience for the 1999 Caribbean Youth Science Forum.

Members of the National delegation to the 40th International Youth Science Forum 1998 served on the planning committee of CYSF and acted as hosts.

With political independence, industrial development and a greater understanding of the relationship between socio-economic development and scientific and technological competence, there is growing recognition in the region that our human resources must be trained to keep pace with scientific and technological development. This is essential if the region hopes to survive and prosper in an age of liberalisation and industrial competitiveness. It has become imperative, therefore, that Caribbean countries develop a culture and social attitude in which science and technology is an integral part, as well as, a culture of innovation, the prerequisites of which include continuous learning, problem solving and the creation of new knowledge.

Outside of the formal education system, science popularisation activities play a vital role in achieving the objective. Science popularisation activities require financial resources, a commodity that is scarce in many Caribbean countries.

The National Institute of Higher Education, Research, Science and Technology (NIHERST) founded the NIHERST/NGC National Science Centre with funding from the National Gas Company of Trinidad and Tobago. The centre was formally opened on 1 February 1998, using 16,000 square feet of rented warehouse space. Prior to its opening, it existed, for about six years, as a travelling exhibition called Yapollo (an Amerindian word meaning to discover) using a container which housed thirty interactive science exhibits, all built locally. The exhibition also included a star lab. The Yapollo Science Exhibition was shown at a number of venues including school halls and shopping malls in Trinidad and Tobago. With the formal opening of the science centre, Yapollo finally obtained a much-needed home that could welcome the population, not only on special occasions, but continually.

International agencies have played a key role in initiating and supporting science popularisation programmes in the region. NIHERST, as the Caribbean leader in science and technology popularisation activities, applied for and received financial support for funding a number of science popularisation projects. This paper addresses one of these activities that was developed and executed as an outreach activity of the National Science Centre.

Caribbean Youth Science Forum – 1999

This project was fully funded by the Organization of American States (OAS). Experts in the field of Science and Technology along with 124 students representing 12 countries in the Region met at the first-ever fully funded Caribbean Youth Science Forum (CYSF) from 2-6 August, 1999. The Forum hosted by NIHERST was held at the University of the West Indies, St. Augustine campus, Trinidad and Tobago.

Countries represented were Antigua, Bahamas, Barbados, Dominica, Grenada, Guyana, Jamaica, St. Kitts, St. Vincent, Suriname (five participants each), St. Lucia (seven participants) and Trinidad and Tobago (67 participants). Members of the national delegation to the 40th International London Youth Science Forum 1998 served on the planning committee of CYSF and acted as hosts.

Forum objectives

The Forum sought to:

- Develop innovative informal approaches and strategies to strengthen Science and Technology education in the Region
- Foster co-operation among Caribbean countries in Science and Technology popularisation

- ♦ Encourage the pursuit of careers in Science and Technology
- ♦ Promote interaction with regional and international scientists.

Forum programme

In order to achieve the Forum objectives, the following activities were planned and executed:

- Plenary lectures

The Joy of Discovery and the Power of Scientific and Technological Knowledge to Reform the World around us
Dr Eric Jolly

Dr Eric Jolly

In his lecture, Dr Jolly had this to say "What is the joy of discovery? For me, it is the recognition of the power that you have to control your own life and influence the world around you, to improve the communities you live in and to give back to society. It is the liberation from ideas, from technologies and science that allow us to improve our world."

This quote was taken from Dr Jolly's opening remarks. He developed this by building his presentation around the prelude to Donald McCarto's book entitled "Literacies Of Power". He stated: "...the next literacy of power, the next literacy of control and the next literacy of influence is that of science and technology. As you come here today, you are beginning a journey in which you are developing your leadership for this new literacy. A literacy in which all communities must be versed".

Dr Jolly stated and cited how technology is used as a powerful tool to regulate the boundaries between people since it can be designed to include or exclude. He noted that these boundaries are controlled by those who are skilled in scientific and technological fields. In this regard, drawing on his experience as a person of Native American origins and having worked with minorities, he stressed that learning and meaning should not be individualistic. Contextual learning according to Dr Jolly, is powerful because it extends the capacity of the learner to influence their world, to integrate new knowledge and to contribute to the collective wisdom of the community.

He concluded that our goal today is not to worship science and technology, but to tame it and bring it into the lives of our community, to find a way to give it meaning. As such, it is important to break the boundaries of formal education by creating community programmes which enliven people with the knowledge that they already are scientists, mathematicians and technologists. Against the background, he called upon participants to exercise their power by asking questions to form the issues of science, to inform the direction and development of science and to create debate.

Dr Jeffrey Dellimore

The Business of Science
Dr Jeffrey Dellimore

One of the concerns in the Caribbean is that many of our outstanding students are not selecting science for study beyond the form three level. Business subjects are the major competitors. Two main reasons are offered:

- It is easier to gain good grades in business subjects
- Business graduates earn much better salaries that science graduates

In addressing this concern, Dr Dellimore stated that:

- The essence of science is more than application of the "scientific method". It is also an attitude that creates a relentless search for new knowledge and understanding.
- To enhance knowledge capital, and hence competitiveness of Caribbean businesses, we need to first establish a scientific culture within our business enterprises.
- Scientists must accept responsibility for the 'scientification' of business by:
    - Popularising scientific issues/problems
    - Helping to demystify the scientific basis of business processes
    - Not waiting on others to create jobs for scientists in business but by creating science jobs and thereby demonstrate the value of scientific

knowledge and skills to existing enterprises or create new science-based enterprises themselves.

The Passage of Nuclear Waste through the Caribbean and its possible effects on Ecotourism
Dr John Agard, BSc, MSc, PhD

In the late 1970s, France agreed to treat 2.8 tonnes of Japan's nuclear waste based on the condition that it would be sent back to Japan to be stored. Dr Agard examined the path of the shipment of reprocessed plutonium and uranium oxide fuel from Europe back to Japan via the Caribbean Sea. He concluded that the transport of nuclear waste through the region poses a threat to the sustainable development of eco-tourism. This is a significant challenge since eco-tourism is the fastest growing tourism sub-sector in the Caribbean.

AIDS – Research and Implications
Dr Jeffrey Edwards

Dr Jeffrey Edwards

Dr Edwards presented data from current HIV research studies conducted in Trinidad and Tobago. Data presented include the fact that the first case of AIDS in Trinidad and Tobago was diagnosed in 1983 in an homosexual male. During 1983-1984 studies were conducted on 100 gay/bisexual males attending the clinic for treatment of sexually transmitted diseases. The first cases of AIDS in women and children were diagnosed in 1985. Since then, data shows a transition to a rapid predominantly heterosexual HIV/AIDS epidemic. In 1996, heterosexual males and females accounted for 56.6% of AIDS cases.

All presentations were followed by discussions in which students fully participated.

- Concurrent sessions

Problem Solving in Science and Life
Rev. Steve West, BSc, MSc

Rev. West presented a paper which traced the need for problem-solving skills from the development of primitive man to learning in everyday life. He explored the use of problem solving techniques in the teaching of Science and outlined three models namely: Creative Problem Solving (CPS) by Noller and Mauthe

(1977), The Big Six Skills Model of Eisenberg and Berkowitz (1966) and A Thru E Approach to Problem Solving by Woodcock (1999).

Caribbean Disasters
Ms Joan Latchman, BSc (Hons), M.Phil.

This lecture dealt with the monitoring and detection of seismicity and volcanism in the Caribbean. Located on the eastern boundary of the Caribbean Plate, the islands of the Eastern Caribbean are subjected to subduction of the Atlantic Ocean lithosphere beneath the Caribbean Plate. This results in seismicity and volcanism. However, Ms Latchman illustrated that once the nature of the hazard is understood, steps can be taken to reduce the risk.

Natural Medicine – Alternative Medicine
Dr Harry Ramnarine MBBS.

In Trinidad, Dr Ramnarine is a pioneer in the field of energetic medicine. He is currently working in the field of natural and alternative methods of healing as well as conducting research in this area.

Dr Ramnarine provided participants with an overview of his work and research. He complemented his presentation with practical demonstrations which included a diagnostic procedure using the bioenergetic system of the body and the Vegatest Method by Helmut Schimmel.

Dr Norma Andrews

Dr Andrews introduced participants to the underlying scientific basis of alternative medical practice and discussed the importance of using systems that adhere to these scientific principles. She touched on the range of options available in alternative medicine and pointed to sources from which further information could be accessed.

- Field trips

These allowed participants to see research science and technology in action and observe nature centres. The aim here was to remind participants that industries and nature centres must be able to co-exist. The trips included industries such as Angostura Limited and KC Confectionery Ltd. and nature centres such as the Wild Fowl Trust and Asa Wright Nature Centre and a research institution.

Angostura Limited
Situated in Laventille, just east of Port of Spain, Angostura Limited is home to the world renowned Angostura aromatic bitters and Angostura-Barcant

Butterfly collection. The 170-year-old secret bitters blend was developed by Prussian doctor, Johann Gottlieb Benjamin Siegert. Today, in addition to aromatic bitters, Angostura produces a variety of alcoholic beverages including rum.

The CYSF tour included:

- Angostura-Barcant Butterfly collection
- Aromatic bitters production facility
- The bottling line and rum distillery

Asa Wright Nature Centre
The Asa Wright Nature Centre, 1,200 feet up in the rainforest of the Northern Range, is an old estate house that has become a conservation and study centre for naturalists and bird-watchers. This estate is home to over 170 species of birds as well as the most accessible colony of oilbirds in the world.

The Sugar Cane Feeds Centre
The Sugar Cane Feeds Centre is an institution of applied research, demonstration, development and training tropical livestock production. Located on 60 hectares of land at Longdenville, Central Trinidad, the Centre promotes integrated production systems which use farm and other local feed resources.

- Students' presentations

In preparation for CYSF, the following topics were sent to the participating countries:

- Pollution and its effects on the environment
- Food technology
- The Impact of tourism on science and technology popularisation
- Natural disaster preparedness
- Biotechnology
- Biological Diversity
- Information Technology.

Student presentations at CYSF

Working in teams of three to five persons, students were required to:

- Select a topic and identify a need/problem associated with the topic

- Design and develop an innovation that will satisfy the need or solve the problem identified
- Bring to the forum full reports including suitable products where possible.

Many groups addressed environmental issues and offered feasible solutions to the problems identified. Some of the solutions were:

- Educating the public about the damage that pollution does to the environment.
- Enforcing laws that govern pollution.
- Making adaptations to a reef buoy designed by John and Harold Hudson of the Key Largo National Marine Sanctuary, Florida. A group from Tobago worked on the adaptations in an attempt to prevent biotic and antibiotic damage to the Buccoo Reef.

St. Vincent addressed "Natural disaster preparedness" and suggested the development of a mobile sea wall that would protect the southern region of the island. The wall would be controlled by a calibrated vibration sensor or hydrophone miles away from land. When it senses abnormal wave motion, it transmits the information to a latch circuit that raises the seawall. Instead of having a water build-up, the waves would refract outwards, away from the island.

Dominica also looked at National disaster preparedness and suggested Public education, training in Shelter Management and communication skills.

The Barbados team examined "Sustainable biodiversity with special reference to saving the turtles in Barbados through the hands of children (6-10 age group)".

This team felt that if the country planned to save the turtles generation after generation, it needed to target the children. The team collected all the data on the turtle and produced an activity booklet and a story booklet designed to disseminate information about turtles to children.

A Trinidad and Tobago group designed a Bi-O-Carrier which simultaneously maintains food at two different temperatures. The innovation comprises two compartments, one hot and one cold.

Summary of Evaluation

Findings of the evaluation on the experiences and views of participants included:

- Duration: Almost all participants felt that the Forum should have been conducted over a two-week period. One week was too compact for the programme of activities. Over 90% felt that it should be an annual event.

- Concurrent Sessions (lectures, student presentations): Participants felt that concurrent sessions should be omitted. They felt it was difficult to choose which they should attend and that they therefore missed interesting sessions and valuable information.

Delegates at the CYSF, 1999

- About 89% of the participants said that some of the sessions helped them to select career goals.
- Over 95% said that they had the opportunity to receive answers to burning issues in science and technology.
- There was a very positive response by participants, when asked about the usefulness of the forum in providing information about new careers. Forty-eight per cent agreed that the information was useful, 38% strongly agreed. This gave an aggregate of 86%.

A presentation from a member of one of the 12 Caribbean countries represented at the CYSF.

- Participants indicated that there was a good rapport establish between the Lecturers and the Participants, so that burning questions and relevant issues of Science and Technology were aired and discussed. The response rate for this option of the question was 95%.

Some direct comments were:

Grenada
"An interesting, enlightening, exquisitely organised forum which I have benefited greatly from. I wish to see it continue on an annual basis."

Guyana
"It is commendable that there were so many competent presenters. The information they passed on was invaluable. Interests in various subjects were stimulated in a way that they never have been before."

Tobago
"This project was a well thought out one and by all means it should be a continuous even so that other students can benefit like I did".

St. Kitts

"I strongly think that this forum should be held every year as it would give many more young people the opportunity to be a part of such a momentous occasion. I am proud to say that I attended the first CYSF and I am already looking forward to the one next year."

Althea Maund *is Science Educator at the National Science Centre (NIHERST), 20 Victoria Avenue, Port of Spain, Trinidad, West Indies.*

Email: yapollo@tstt.net.tt

Living With the River: The Rideau River Biodiversity Project

Monty Reid
Canadian Museum of Nature, Ottawa, Ontario, Canada

The Rideau River winds out of a series of lakes that occupy an old fault in the landscape of southeastern Ontario. From the town of Smiths Falls to the bustling city of Ottawa, it drains an area of almost 4,000 square kilometres. For centuries the River was an important food source and transportation route and, with the construction of the Rideau Canal in the 1820s, its significance increased. Today, it remains in use as a transportation link and is a highly prized recreational river.

Along with its marshes, lakes and forests, the Rideau is a vital natural habitat for a surprising variety of plants and animals. And it is, like most natural areas, under increasing stress. Whether from the proliferation of zebra mussels or the increased use of pesticides along its shores, the river faces a swarm of challenges.

The Canadian Museum of Nature, Canada's first and largest natural history museum, has had a longstanding interest in the Rideau. In 1998, together with the Rideau Valley Conservation Authority, the Museum initiated the Rideau River Biodiversity Project to study the long-term health of the river. Funded in part by a grant from Montreal's EJLB Foundation, the three-year project was designed to combine the expertise of many scientists in providing an in-depth report card on the River's health. This would lead to an exploration of how local needs could be reconciled with sustainable management of the River's biodiversity. The multi-disciplinary approach remains one of the Project's important features. Zebra mussels or snapping turtle populations are not being studied in isolation. Instead, Museum scientists are attempting to study all aspects of the Rideau's biosystem, from water chemistry to migratory birds.

It is an ambitious approach, and one that has had Museum researchers and many volunteers wading, diving and boating all along the River. Already, some interesting findings have appeared. For instance, the aquatic flora of the Rideau seems to be surprisingly rich, with 55 species identified to date. The world's smallest flowering plant, *Wolffia*, is alive and well in the River. Several fish never before seen in the Rideau turned up early in the Project. One, *a Freshwater Drum*, seems to have moved in from neighbouring waterways. Another, an Oscar, is more at home in the Amazon basin and is probably an escapee from an aquarium.

Also reassuring is the discovery that populations of frogs and turtles do not show a high incidence of abnormalities. This is possibly connected to the discovery that levels of pollutants due to fertiliser use are also in decline. Less happily, the expansion of the zebra mussel population was also confirmed.

A second crucial element of the Rideau River Biodiversity Project is community involvement. For centuries, settlements along the Rideau have depended on the River for their own vitality. Over that time, highly detailed stores of local knowledge have been built. In an effort to utilise that knowledge, and to ensure that the Project's studies were relevant to the river communities, the Museum sought their involvement. Community groups helped to identify areas of concern and continue to participate in the data gathering. Research results and recommendations will be turned over to the same groups, where they will be used to make informed decisions on issues affecting the Rideau's biodiversity.

Governmental organisations such as the Regional Municipality of Ottawa-Carleton and Environment Canada have been important partners in the Project. Community advisory groups made up of volunteers review the Project and provide direction. Their suggestions led to the expansion of the research area to include the lakes at the upper end of the watershed. But even more important are the area-based organisations such as the Rideau Environmental Action League, based in the town of Smiths Falls near the headwaters of the River. This volunteer organisation has a ten-year track record of successful initiatives, from radio shows to water quality seminars, and continues to be an important support to the Biodiversity Project.

The Project is not just about data collection and the subsequent research. It includes a strong public education component too. This takes many forms, including publications and a small travelling exhibit, but it includes more innovative activities as well. Identification workshops took museum scientists to many river communities to assist in the identification of local species. A field guide to species along the Rideau is in production. Boat tours, led by Museum researchers, provided a careful but detailed look at environmentally-sensitive areas. A locally-produced television series focusing on communities along the River was produced and broadcast, and a half-hour video was also prepared. A Turtle Hot Line was set up in order to obtain as much information as possible about turtle species living and breeding along the Rideau. So many calls came in the first year that they swamped the voicemail capabilities of scientist Mike Rankin.

The project is now in its third year, and while the research continues, the major focus of the final year is to set up an independent, community-run organisation to continue the work. To that end, a 30-member community roundtable has been convened and a detailed plan titled RiverCare 2000 has been developed to identify required work to preserve and sustain the river. It includes everything from new sewage treatment plants to shoreline restabilisation.

The Rideau River Biodiversity Project has proven to be an important application of Museum research to an issue of immediate relevance to the local community. It has not only enhanced knowledge of the life of the River, but it has provided solid data upon which political decisions must be made. It has involved many museum disciplines and it has been the catalyst for a lively and sustained interaction between the Canadian Museum of Nature and its immediate community.

Monty Reid *is the Manager, Exhibition Services, Canadian Museum of Nature, P O Box 3445, Station D Ottawa, Ontario K1P6P4, Canada.*

Email: mreid@mus-nature.ca

Science on the Move – Exhibit Design Workshops

Allen Rooney
Questacon – The National Science and Technology Centre, Canberra, Australian Capital Territory, Australia

Background

In 1995, Questacon, Australia's National Science and Technology Centre, toured its hand-on interactive *Fascinating Science* exhibition to Western Samoa, Tonga and Fiji. The stimulus for this programme came from the UNESCO office for the Pacific States and their concern that the Pacific region has little or no access to such programmemes that are taken for granted in many other regions of the world.

Following the success of this programme, Questacon began consultations with AusAID, UNESCO and the Department of Foreign Affairs and Trade to develop a proposal for a significant new programme entitled *Science on the Move*.

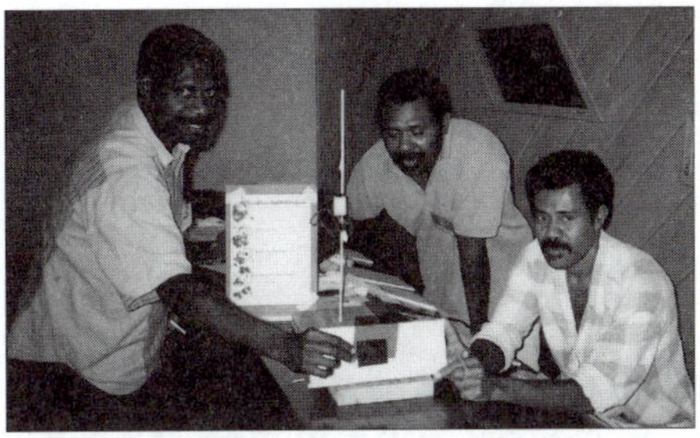

A *Science on the Move* workshop in Vanuatu.

Science on the Move in Fiji.

In May 1996, Questacon worked with educators from Western Samoa, Tonga and Vanuatu to develop the early concepts for this new programmeme. National Coordinators, appointed by participating countries, met during October 1996 to further develop the programme. The exhibition was completed in January 1997 and featured 33 specially designed hands-on exhibits for major population centres and 32 smaller hands-on exhibits for travelling to regional and more remote areas. By August 1997 these exhibits had toured nine Pacific countries – Marshall Islands, Kiribati, Tuvalu, Vanuatu, Fiji, Solomon Islands, Cook Islands, Tonga and Samoa. More than 60,000 visitors experienced the *Science on the Move* exhibition and science shows. Over 300 teachers participated in teacher training workshops designed to introduce interactive approaches to the learning of science. In addition, many local students, teachers-in-training, and curriculum officers were trained as explainers. The *Science on the Move* exhibition provided a highly visible and exciting backdrop which inspired a fresh way of thinking among those engaged in formal and non-formal education in terms of how to communicate science ideas and their relevance to everyday life.

Exhibit design workshops

Participants in Fiji collaborate in an exhibit design workshop.

The next phase of Questacon's *Science on the Move* programme in the Pacific, involved the presentation of two Exhibit Design Workshops aimed at education professionals with an emphasis on those involved in informal education. These workshops took place in Vanuatu and Fiji in May 1998 and were facilitated by staff from Questacon and the Investigator Science and Technology Centre.

Participants included staff from Cultural Centres and Museums as well as science teachers. They were selected according to the following criteria:

- Involvement (or interest in becoming involved) in the design of exhibitions, displays or other types of programmes which seek to educate communities
- Willingness to participate actively in order to generate new ideas for interactive style exhibits that could be built for museums, cultural centres or other informal education institutions
- Interest in science and technology as it relates to Pacific contexts – including traditional contexts

The workshops involved a total of 27 representatives drawn from surrounding countries: Vanuatu, Papua New Guinea, Solomon Islands, Cook Islands, Fiji, Niue, Samoa and Tonga.

The objectives of the workshops were to:

- Promote awareness and understanding of science and technology and their relationship to everyday life
- Increase skills and capabilities of staff in the development of interactive programmes

- Develop an awareness of the potential of interactive exhibits to communicate not only the new, but also the traditional sciences and technologies relevant to Pacific countries and contexts
- Promote and develop a network of non-formal and informal education professionals throughout the Pacific region.

The workshop process

Each workshop began with the participants modelling some of the key stages of the exhibit design process. In small groups they were given a particular science concept and assigned the task of designing an interactive exhibit that would enable visitors to explore that concept.

After discussion and some rough sketches the groups set about building a small prototype of their exhibit using a range of tools and materials provided. Once the prototype exhibits had been constructed, often after significant modifications to the original ideas, the groups prepared text panels, which provided clear instructions on how to operate the exhibit, what to observe and how the exhibit related to science in everyday life. The groups then presented their exhibits and there was time for each group to experience the other exhibits.

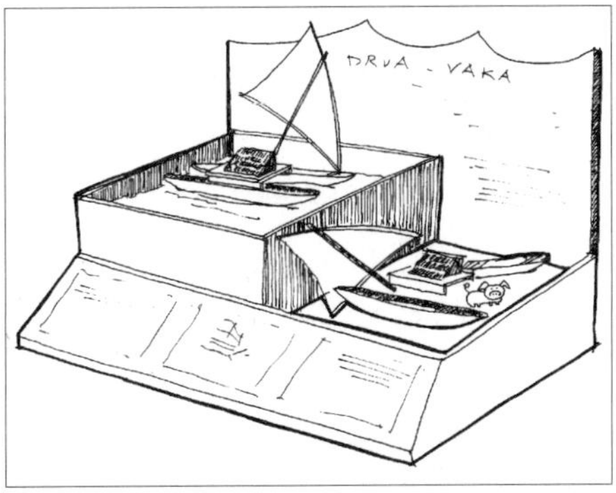

Making it relevant. The peoples of the South pacific region have used the double hull canoe for long distance voyages and fishing for many years. Using the double hull canoe as a model, this exhibit explores concepts of flotation and buoyancy, centre of buoyancy, shape, balance and stability.

This activity provided participants with some practical experience of factors to be considered when designing an interactive exhibit.

A key idea of the workshops was to develop exhibit concepts that would relate to science and technology – especially in traditional contexts – in the lives of people in the Pacific. Participants were asked to list topics that involved science or technology concepts or processes drawn from modern or traditional life in Pacific countries. The list was extensive and included topics such as building techniques and fire making and navigation and communication methods.

In small groups participants chose one of the topics and generated ideas for an interactive exhibit. After discussion the group agreed to certain parameters for the exhibit designs:

- The exhibit concept may be based on a topic with a science and/or technology basis.
- The exhibit concept may focus on either traditional or modern contexts.
- The exhibit concept should be developed so that it is possible for the exhibit to tour from one Pacific country to another.
- The exhibit concept should be one that participants would be genuinely interested in building if funds were provided.

The groups documented their ideas using the following format:

- Exhibit name – what is the title of this exhibit?
- Message – what message(s) does this exhibit communicate?
- How it works – what does a visitor do to operate this exhibit? What does a visitor notice?
- Concept – what science principles and concepts does this exhibit explore?
- Relevance – how is this exhibit relevant to Pacific contexts – either traditional or modern?
- Additional notes – does this exhibit suggest ideas for other exhibits? Are there particular issues that may need to be addressed?

Once an exhibit idea was shaped, the group discussed their design concept with an Australian facilitator who then developed an illustration to show a possible representation of the exhibit. As exhibit concepts were completed they were displayed on the walls and recorded on a computer in readiness for publication.

Participants changed groups from time to time. Different groupings helped to stimulate new ideas and encouraged professional relationships to develop. At various stages during the workshops the groups shared brief reports on the ideas they had generated. In total, 47 exhibit design concepts were produced with many suggestions for additional exhibits.

All participants were committed to the importance of communicating aspects of their traditional cultures and realised that their institutions – schools, museums and other cultural centres – had an important role to play. A recurring theme throughout the workshops was the importance of people being actively involved in their learning.

Staff from cultural institutions were keen to explore new ways to communicate information and knowledge contained in their collections of artefacts and saw an interactive approach as an effective way of passing on skills and traditional knowledge.

The teachers in the groups were keen to explore a wider range of interactive activities that could be drawn from the exhibit concepts and included in their school programmes.

All participants gained an increased understanding of the importance of hands-on activities in the education of people of all ages.

Allen Rooney *is the Education Officer at Questacon – The National Science and Technology Centre, King Edward Terrace, Canberra, ACT 2600, Australia.*

Email: arooney@questacon.edu.au

Exhibitry: Contributions to Entertainment and Learning

Geoff Snowdon
Petrosains Discovery Centre, Kuala Lumpur, Malaysia

Petrosains Discovery Centre located in Kuala Lumpur, offers visitors 7,000 square metres of exibitry. The focus is necessarily on oil and gas, since Petrosains is owned and funded by Petronas, Malaysia's Oil and Gas Company, 'a multinational oil and gas company of choice'.

Petrosains Discovery Centre, Galeri Petronas and the Dewan Filharmonic Concert Hall and Orchestra, form an exciting trilogy of cultural offerings that is science, art and music made available to all Malaysians by the Kuala Lumpur City Complex (KLCC) project. KLCC includes Petronas Twin Towers, a magnificently modern shopping centre complex – SURIA – and a grand green parkland, itself a spectacular and very popular feature in the heartland of Kuala Lumpur. Petrosains has premier location in a very modern setting.

Demonstrations intrigue adult visitors to Petrosains Discovery Centre.

Petrosains has other uniqueness in that people enter from the shopping centre by way of a 4.5-minute ride in a car similar to a Disney quiet ride experience. This prepares visitors for more than a two-hour journey that is an entertaining and learning experience through exhibitry with its focus on the science and technology of the oil and gas industry. This has been interpreted strongly in the exhibits and the theatrical settings that form the mainstream storyline. Since opening on 25 March 1999, and even for some months leading up to the opening, staff have introduced an enhancement programme that has seen another layer of exhibitry introduced to add value to the visitor experience. The enhancements have resulted in purchase of a 200 square metre exhibition for 4-8 year olds. The addition of exhibits that provide some lead in science concepts aim to assist the visitors in their explorations and discoveries, as well as a wide range of performance oriented exhibitry aimed at engaging visitors in fun learning and entertainment.

One of the vision statements of Petrosains is 'to be a dynamic world class science discovery centre that inspires fun learning, creativity and innovation'. This is rapidly being the accepted vision for Petrosains. Staff are planning, arranging, adapting to, interpreting, understanding and appreciating the scope and the potential, inherent in this vision statement. There is a keenness to work towards reaching this vision that this moment is energising and providing definite direction and stimulation for staff in Petrosains. These are exciting times for a centre that is still one of the newest centres in Asia.

The concept 'exhibitry' is interpreted in Petrosains to include anything that is provided to add value to the visitor experience. The main 'exhibitry' focus is the mainstream story line, the hardware for which includes:

- over 150 interactive exhibits
- many information panels
- theatrical settings including an impressive diorama depicting early geological time (200 million years ago)
- two helicopter simulators that take visitors to a large scaled oil platform experience
- three computer multimedia discovery stations
- a virtual reality molecule experience
- a communication device that is uniquely a hand held computer that provides a next layer of information downloaded by radio signals by request of the user and which can be stored for retrieval later on exiting the exhibition area.

Exhibitry includes enhancements, many of these displayed on special platform stands. We have introduced an earthquake machine experience, a Formula 1 driving simulator, a 'Pit-Stop' exhibit that allows visitors to compete to change tyres, as in Formula 1 racing teams and a very popular small 'Digger' that shows 'simple' machine principles for people of all ages. Many of these enhancements are added to increase the 'fun learning' of the Petrosains experience.

There is a software side to our exhibitry. This is offered through the enterprise and creativity of staff. Our front of house staff includes Visitor Services staff along with staff from our Exhibitry Department. This Department provides facilitators and the exhibit development and maintenance staff. A volunteers' programme supports both Departments. It is the software side of exhibitry that contributes significantly to the value adding visitor experience. Visitor Services staff assume responsibility for overall comfort of a visit while the Facilitators provide the engagements and the performance that are key to enhancing and providing the fun learning, entertaining parts of the visitor experience.

Facilitators build constantly on their repertoire of skills for engaging with visitors. Our approach is through Constructivism. This we work on and are developing better ways to monitor pragmatically how we use and bring our particular style and attributes to interpreting Constructivism in our Centre.

Facilitators are training and presenting science shows. We are developing an approach based on mini shows – engagements that are five-minute vignettes. Facilitators are building confidence in 'busking' science – impromptu engagements frequently using props from our shop. We include puppet shows in our Science Play exhibit area and are keen to explore other ways theatre might be employed. We are considering regular themes for our children's area. We have a need to be creative and innovative in our thinking so that return visitors will know our centre is dynamic in various ways.

Children visiting the Petrosains Discovery Centre gather around for some fun.

We are mindful that we need to shore up our links with scientists, with relevant Ministries responsible for education, science and tourism. This is so we can make sure their needs and their reach is included in software programs we create for use with various target groups of visitors. The reach of S & T in the oil and gas industry we interpret quite liberally. There are few parts of science and technology that do not touch the industry in some way. It is pervades our lives and cultures. In this approach we can contribute significantly in the assistance of teachers, students, and professionals in the science and technology industry and especially in the science communication industry. We look forward to internships, student, teachers, scientists, writers and others as professionals in residence. These things we see as part of the dynamic world class offerings that are possible and appropriate for science centres as well as for us to achieve our vision. We are keen to entertain staff exchanges as we see these as win-win sharing situations and the opportunity for our staff to grow through such sharing.

We are in our second year of operations. We targeted 200,000 visitors in our first year. This we did in eight months. Our location in SURIA, which attracts over 400,000 weekly, helps us be confident we can reach more people in our second year. We build our confidence for increased visitation on many factors – in our approach to exhibitry in all its dimensions as well as from the Customer Service Index survey, we have recently completed where visitor's responses have shown emphatically our approach is appreciated.

Staff, who really had no experience of the science centre industry when they were appointed, are impressively seeing and accepting the many opportunities there are to help them grow. These are exciting times at Petrosains. Do introduce yourself when you visit. We will be keen to have your feedback on our exhibitry and keen also for you to share your ideas and successes with us.

Thus far it 'feels good'.

Geoff Snowdon *is the Consultant to Science Museums, Secretary ASPAC, AED Petrosains, P.O. Box 124444, 50778, Kuala Lumpur, Malaysia.*

Email: gsnowdon@tm.net.my

Section 2

Making the Most of a Visit

School excursions have been popular for many years, but they should be more than just an excuse to get out of the classroom. To make a visit to a science centre or museum a learning experience for the students, both teachers and centres need to be prepared to maximise the visit to transform an excursion from "an escape from school" into a worthwhile informal learning experience.

The articles in this section establish a link between the formal and informal education settings to enable both teachers and science centres to make the most of the interaction. For the visit to be valuable the visitors (i.e. teachers and students) need to be prepared. The science centre or museum must also be able to offer an experience that will allow this to happen.

The first article by *Leonie Rennie and Terry McClafferty* serves as a guide to teachers taking their students on trips to science centres, museums, zoos, botanical gardens and aquaria. The article outlines necessary steps before, during and after the visit, enabling all parties to have an optimal experience.

Judith Arrowood and Hooley McLaughlin describe how they have set up an education programme within the Ontario Science Centre that can benefit not only the students visiting, but also teachers. This theme is carried over into the following article, by *Neville Gardner*, with reference to a specific programme that makes available, to both students and teachers, resources not often available in schools.

Brett Dunlop's article looks at the museum's perspective of taking into account clients' needs during the design process of an exhibition. The final two articles discuss the importance of evaluation. *Robin Garnett* examines the pre-visit needs and the post-visit evaluation of teachers in relation to satisfying their expectations and objectives of visiting the Science Centre at the Manawatu Museum and Art Gallery. The final article of this section, by *Paul McCory and Melanie Quin* examines the outcome of student visits to Techniquest and how these are integrated into normal classroom work.

Close collaboration between the formal and informal education sectors, with emphasis on curriculum-based activities, can make a visit to a science centre or museum not only a pleasurable experience, but one that will complement and enhance classroom activities.

Visiting a Science Centre or Museum? Make it a REAL Educational Experience!

Léonie J. Rennie
Curtin University of Technology, Bentley, Western Australia, Australia

Terence P. McClafferty
Western Australia Museum, Perth, Western Australia, Australia

People visit science centres, museums, zoos, botanical gardens and aquaria for many reasons. When teachers take their class for a visit, there is usually an educational reason, if only so that the visit can obtain the official stamp of approval! But it takes time and effort to organise the visit, so how can teachers be sure the visit is worth the effort? In the following, we present some guidelines we have distilled from an extensive review of research evidence. Most of the guidelines are common sense, but it is surprising how few teachers seem to realise that they really do make a difference. The best educational outcomes are a result of good preparation, a focused but flexible visit experience, and some follow-up activity back at school.

Before the Visit

Why Are You Going?

This is not a trivial question! The reasons why teachers take their classes to science centres or museums determine how they should prepare themselves and their students to maximise the educational outcomes. If the purpose of the visit is to stimulate or motivate students, then the aim is to arouse interest and curiosity about concepts or ideas that the students might be finding rather mundane at school. The choice of exhibits will relate to school work, but provide new (and perhaps extra-curricular) perspectives on those concepts. Alternatively, pre-visit discussion at school can help students to come up with a list of questions to be investigated at the museum during the visit. If the purpose of the visit is to introduce a new topic, then exhibits will be selected to demonstrate a variety of concepts to be covered in the topic. Thus, students will leave the centre with a range of unanswered questions to pursue back at school. If the visit is to revise and consolidate the learning of concepts, exhibits should be chosen which provide new demonstrations of related phenomena and applications of associated properties. The overall aim is to transfer the enjoyment and enthusiasm aroused by the students' visit to the achievement of science objectives back at school.

How long should the visit be?

Depending on the age of the students, a visit might be between one and three hours. If they have not visited before, students need time to get their bearings, time to do serious exhibit-oriented work, and time to do the things they want to do, such as visit the shop and look at other galleries not on the teacher's list. But working with exhibits requires concentration and "museum fatigue" will set in. The length of the visit needs to strike a balance between these things.

Teacher Preparation

Know what you are going to find when you get there. The best way is to visit yourself. When you know what exhibits are there, you can work out what concepts or phenomena they demonstrate, what level of thought processes they require to be understood, and whether there are worksheets or other cues, like labels, available to help understand the exhibits. When you have this information, you can make the visit fit into your teaching programme. Most science centres and museums have education officers who can help in planning a visit. Often there are informative materials and related activities designed to make the teacher's job easier. Make use of them.

Student Preparation

Find out whether students have visited the science centre or museum before. If they have not, then novelty may be a distracting factor in the visit. You can help with some advance information about basic things, like a map of the centre, showing where to meet, lunch areas, toilets and the shop. Students will probably be more concerned about whether they will be able to visit the museum shop than the teachers' plans for the visit! Get their help in planning the visit. If you have a specific set of science concepts in mind, they need to have the necessary background knowledge and skills to use and understand how the exhibits work. This might require some pre-visit instruction. If your objectives are more general, you can encourage students to ask their own questions and to find out the answers through their own research during the visit. Make sure they do not have too many questions, and that finding the answers to them is a realistic task for the science centre or museum. The important point is that if students know what learning is expected of them, they can be more self-directed in achieving it.

During the Visit

Getting Started

If this is their first visit, students will need some time to settle down. Expect some playing and exploration with the exhibits which might look more like fun than work, even when students are seriously working!

Working with Exhibits

Teachers can help students to keep track of time and their learning objectives, as well as being on hand to make suggestions to extend their thinking and understanding. Obviously, students with different levels of skills or preferred styles of working, may need different kinds of help. Attendants and parents, or others who chaperone school groups, are other important sources of cues to help students understand the exhibits. Make sure helpers know that the purpose is to open-up students' thinking rather than tell them to the right answer.

Working Together

Students enjoy the social interaction with their friends that visits afford. Make the most of it. Encourage students to work in small groups and share the responsibilities associated with learning. Research shows that a great deal of peer teaching occurs during visits, with children asking each other questions, reading labels aloud, and showing each other how the exhibits work.

Keeping a Record

It is beneficial for students to keep a record of what they discover, so think about the best way to do this. If there are open ended-questions for students to find answers to, then an open-ended kind of record is needed. Digital cameras allow students to take images back to the classroom to incorporate in reports or other displays. A word on worksheets: Don't let them become "treasure hunts", with students dashing about looking for an exhibit to find out some fact or other. Good worksheets relate directly to the exhibits themselves, not to their labels, and promote opportunities for meaningful, cooperative group learning. This means that one worksheet per group can be effective, rather than one each, which encourages trading answers. Some teachers help students to prepare their own worksheets for visits to zoos and museums.

Finishing Off

Check how students are progressing near the end of the visit, so the remainder of their time can be structured effectively. Do not forget that students have their own things they want to do, so let them have at least 20 minutes to explore the exhibits that interest them.

Back at School

Common sense suggests that post-visit activities should reflect the varied nature of the experiences students had at the science centre or museum. Young children, in particular, should be given the opportunity to share their experiences and findings with their peers through class presentations, group reports or posters. Students can plan further research or experiments based on what they have found out. In subsequent lessons every opportunity should be taken to refer back to exhibits and activities experienced during the visit, thus reinforcing and extending the learning which occurred.

Visits to science centres and museums can promote students' engagement in school science. They may find the visits more interesting and enjoyable than effective class lessons and this is a strong reason to capitalise on their interest when back at school. The best way is to integrate those visits into the teaching programme in ways that complement the learning activities at school.

Léonie Rennie *is Professor of Science and Technology Education at the Science and Mathematics Education Centre, Curtin University of Technology, Bentley, Western Australia 6102, Australia*

Email: l.rennie@smec.curtin.edu.au.

Terence McClafferty *is the Manager of the Discovery Centre at the Western Australian Museum, Perth, Western Australia 6000, Australia.*

Email: tmcclaff@museum.wa.gov.au

The Unique Role of Science Centres in Immersing Students and Teachers in Real-World Science and Technology

Judith Arrowood and Hooley McLaughlin
Ontario Science Centre, Toronto, Ontario, Canada

Background

The Ontario Science Centre, an agency of the Ontario Government, is located in the large Canadian multicultural city of Toronto. Since its opening in 1969, the Centre has served more than 1 million visitors a year with a variety of experiential exhibits, interactive public programmes, as well as a unique selection of educational programmes.

The 12,000 square metres of public exhibition space cover topics ranging from physics, chemistry, and biology, to the Internet, psychology, cultural studies, sports, engineering, space technology, and environmental studies. Inseparable from this public exhibition presence are a number of educational programmes aimed specifically at students and teachers working within the formal provincial school system. These include special short programmes for class visits, teachers' guides to the exhibits, half-day-long and day-long classes in subjects such as DNA finger-printing, class experiences in the Challenger Learning Centre (a simulated space science laboratory), and even an in-house school for students in their final year of secondary school.

OSCLUB – Launching a New Programme

Enhancing the regular school curriculum, Centre educational programmes inspire teachers and students to think about their studies from a somewhat more lateral viewpoint. Cross-disciplinary associations and engaging demonstrations characterise a class visit to the Centre. Recently, the Ontario Science Centre has taken this approach a step further. Responding to an invitation from the internationally-based Lucent Technologies Foundation to submit proposals for educational programmes, Pamela Kay, Director of Science Education, and Dr Allan Busch, Senior Advisor for Product Innovation and Development, led the successful grant application for a new programme.

Launched in May 2000, 15 regular and pre-service teachers, and 60 students who have just graduated from grade 9, are participating in OSCLUB. Conscious of an increasing interest from the college/university, business and industrial sectors, OSCLUB has the objective of working together with these groups to find engaging new methods for immersing students and teachers in real-world science, technology and mathematics. OSCLUB is formulated around a partnership between the Ontario Science Centre and the Toronto District School Board in collaboration with Shad International, the Ontario Institute for Studies in Education at the University of Toronto, the York-Seneca Institute for Science, Technology and Education, and Lucent Technologies.

OSCLUB consists of a four-part intensive programme. In the first part, begun in May 2000, sessions have been designed specifically for teachers, to prepare them for work in the Centre environment and for the development of relationships with the research and industrial partner associations. In August, teachers began to work with the students to develop small teams with common interests for projects that will come to fruition by October. At that time OSCLUB participants work on real-world problems, challenges and product development in close association with external research, business and industrial partners. At this point it is anticipated that the interests of the students and teachers in OSCLUB will lead to extending partnership associations with experts working in areas such as biotechnology, aerospace, engineering, telecommunications and global environmental monitoring. Projects may range widely in subject matter, from the development of exhibits on the latest discoveries in biotechnology, for example, to the creation of a computer game on telecommunications for the Lucent Technologies website.

The intent is to allow as much freedom as possible while creating a connection to the real-world of the *business* of science and technology. Finally, during the following July in 2001, at the same time as the next group of students and teachers are beginning their programme, the year-2000 students will return to the Ontario Science Centre to gain work experience, in the Summer Camp programme, or in the exhibit halls where they will host visitors to the Centre.

Collaborating with Teachers

Why does this particular educational programme warrant special attention? One of the emphases of OSCLUB is reaching teachers with the goal of achieving long-term changes to the Centre's relationship with the school system. Working together with teachers specifically has the potential of influencing far more students than is possible through normal class visits to the Centre. Being conscious of the fact that a visit to the Centre, while anticipated with delight by both teacher and student, is often thought of as an alternative approach not connected to, nor reachable within, the typical daily classroom routine, it is gratifying to imagine teachers, once initiated into the OSCLUB approach, carrying this information with them throughout the school year.

A further benefit lies in the decision to include both experienced and pre-service teachers. The inevitable alliances that follow a programme of this nature will lead to the development of coaching networks and mentoring relationships, as the teachers pursue their regular work in the school community. Ultimately, many more students than can be accommodated in the OSCLUB programme itself will be reached through this Ontario Science Centre initiative.

Science Centre Technology

What is special about the science centre setting for OSCLUB? The Ontario Science Centre is large enough to have within it several communities of staff, all professionals who create inspirational science and technology experiences, while focusing on different needs. OSCLUB has broken new ground here as well. The rich partnerships with outside industry and research groups are mirrored inside the Centre by a cross-departmental partnerships. Staff from the Educational Programme, whose background and activities are centred on the education of school students, have provided opportunities for the involvement of staff from the Visitor Experience branch, who have up to now been primarily focussed on creating interactive exhibits and programmes for the general public.

Secondary students work with Ontario Science Centre staff to explore real-world science and technology, using engaging interactive prototypes

The student/teacher interactions being explored in OSCLUB emphasise a quality peculiar to science centres. For a long time now, institutions focussed on the public understanding of science and technology have been developing their own form of technology. In an environment nurtured by a very real connection to the research world though interaction with mentors from the outside industrial, academic and business communities, the OSCLUB teaching material utilises the unique science centre approach of:

- mining the subject matter for that one particularly stimulating piece of information that suddenly can inspire interest for material often considered dry by secondary-school-age students
- allowing students to pursue projects that conjoin apparently unrelated disciplines, and that utilise the experience of experts studying a problem from many different points of view

♦ constructing quirky, fun exhibit prototypes, or developing novel group experiences, to make an explanation clear and communicative.

Ontario Science Centre staff provide opportunities for public and school visitors to *play and learn*. The science centre visit is a time for releasing the brain, for stretching a single concept to its logical, and often amusing, extremes. Chemical engineering procedures, for example the separation of liquids based on their densities, can turn into dramatic, apparently precarious, physical experiments where glasses of fluids are spun around overhead on serving trays suspended from cords. Studies on fibre optic communication may lead to home-made telephones using flashlights or novel musical instruments that can be played using nothing but light transmission. The basic materials of our scientific, technological and mathematical world are the tools for discovery. Science Centre staff can do what scientists would do if there were more time to play – and learn even more.

For the students and teachers in OSCLUB, the science centre environment is stimulating. Participants feel free and open to new ideas and ways of doing and thinking about things. Add the opportunity to interact with, and be mentored by, people from industry, business and research facilities and you have a "learning community," a significant step beyond the typical day-long visit to the Centre. OSCLUB teachers and students are immersed in the science centre environment, and they are a part of the larger science and technology network, a network that will increasingly owe a great deal to these particular participants.

Next Steps

The Lucent Technologies sponsorship is in place for three years, until July of 2002. OSCLUB will be monitored by the Educational Program staff at the Ontario Science Centre for the purpose of developing future programmes with schools and the science and technology community. The Centre also sees an opportunity to use OSCLUB as a template for changes in the way we develop our exhibits and programmes for the public in general. Given the investment in time and effort, staff have a rare opportunity in the OSCLUB programme to participate in brainstorming and idea development and to watch prototype creation and building over a relatively long duration. The cross-departmental partnerships within the Centre will undoubtedly have an effect on the culture of the science centre as a whole. Through the student/teacher programme in OSCLUB, staff will be introduced to what is effectively a new kind of learning laboratory. Ideas generated here, in the rich environment of external partnerships and highly charged student activities, will be subjected to rigorous intellectual and physical testing. It is hoped that the external partners will equally benefit from the exercises.

The OSCLUB programme itself will be subject to evaluation by the Ontario Science Centre, the Toronto District School Board, and the Lucent Technologies Foundation. The most obvious outcome will be a model for future school programmes for students and teachers, with the goal of encouraging more students to consider careers that utilise science, technology and mathematics. Perhaps it is not too far-fetched to hope that the science centre mandate to connect the public more directly to the changing world of science and technology will be realised through mentoring programmes like OSCLUB, where students and teachers work together with professionals from industry, research and business – teaching them as much as the programme participants are taught themselves.

Judith Arrowood *is the Associate Director for Educational Programs and* ***Hooley McLaughlin*** *is the Senior Advisor for Science and Technology at the Ontario Science Centre, 770 Don Mills Road, North York, Ontario M3C 1T3, Canada.*

Email: hooley@osc.on.ca

Responding to Teacher Needs: ‘Fossils’, a Hands-on Education in New Zealand

Neville Gardner
The Science Centre, Manawatu Museum and Art Gallery, Palmerston North, New Zealand

The Science Centre, Manawatu Museum and Art Gallery combines visual and hands-on experiences of science and technology, heritage and art under one roof. A staff of 38 mount about 40 exhibitions a year, including touring shows and those that use our own collections and built exhibits. The organisation serves a city of approximately 75,000, but many visitors come from much further afield.

Nearly 20,000 school (mostly primary age) and adult students a year take part in education programmes. The Ministry of Education provides a grant to support such learning experiences outside the classroom. Close links with local educational institutions (Massey University and UCOL, a technical college), the Royal Society of New Zealand and numerous Crown Research Institutes provide access to wide ranging expertise. The same organisations support adult education through lecture series, field trips and so on.

The majority of education programmes are run by one of our staff; a self-lead is option is available for visiting teachers, but is rarely taken up. There are probably several reasons for this. Teachers may bring their class to study a topic they are not familiar with themselves. Museum staff already familiar with an exhibition or programme can greatly enhance student learning. Teachers say they take away new teaching ideas after watching museum educators in action. They also enjoy watching their students learning in a different setting. Sometimes students who struggle in a classroom setting excel in a hands-on science exhibition.

While exhibition-based education programmes are popular, so too are permanent programmes that make use of particular resources that are not readily available in most schools. These include a Starlab planetarium, a purpose built water science area, and museum and art collections. Museum educators develop new programmes in response to teachers direct requests and perceived needs. The ‘Fossils’ programme described below is one such new programme.

The 'Fossils' programme

Through informal discussions with teachers, and as a result of a study carried out by Brian Lewthwaite of the Massey University College of Education, it became clear that many primary school teachers felt that they lacked the skills and/or resources to adequately cover several aspects of the "Making Sense of Planet Earth and Beyond" strand of the National Science curriculum. As a result of this, an education programme called "Fossils" was devised, in consultation with two geology teachers at Massey University (Clel Wallace and Julie Palmer). The programme was an immediate success, and continues to be popular.

School children examine a fossil at close range.

The two-hour programme starts with an investigation into what a fossil is, viewing and handling examples from NZ and the rest of the world. Students then use a specially made piece of equipment, based around a car jack, to turn loose sediments and shells into a plug of fossil-bearing rock under high pressure. This is a very clear and memorable way of experiencing one method of fossil formation.

Students then work in pairs to excavate replica fossils out of sand trays. They go through the whole process of digging: cleaning, drawing, identifying and interpreting. After identification, fossils are placed in the correct positions in a stratigraphic column, after which the real detective work begins. Using the fossils and their own knowledge of organisms in present environments, augmented by information from the educator, students put together a picture of what NZ used to be like at different periods in the past. An actual picture is constructed in a free standing window frame, encouraging the concept of fossils providing a window on the past. At this point, NZ is placed in the broader context of plate tectonics and global change.

Having practised on replica fossils in a hypothetical rock section, it is time for the real thing. Each student excavates a piece of fossil-rich, million-year-old, mudstone from a local coastal site. Previous activities kept the students focused, but from now on total absorption sets in. A single delicately-spined gastropod may take half an hour to extract, other shells and fragments are quicker. The magnifiers and microscope are in constant use, the conversation is totally focused, and every new find sends students scurrying for clues to identification. Students rapidly take on board the new vocabulary of the palaeontologist.

This excavation can easily take an hour, and the session is rounded off by interpreting the finds; what was our site like a million years ago? The programme was actually developed to be about rocks as well as fossils, but students become so absorbed in the fossils, that rocks are only dealt with after specific urging from the teacher. This rarely happens, as teachers become as lost in the adventure of 'real' palaeontology as their students.

What is the attraction of this programme to teachers?

The following observations are based on questionnaires filled out by teachers at the end of each programme. The most obvious reasons are that it fits the science curriculum, and is hands-on, using resources not often available in schools. The programme is good too because it not only tackles some specifics of earth science, it also encourages basic science skills and attitudes. Social skills and critical thinking are integral parts of the programme. Students are the most obvious beneficiaries of such a programme, but museum educators are also mindful of the needs of teachers. Placing their students in the hands of an expert and enthusiastic educator enables teachers to spend time with individuals, or to expand their own knowledge. If teachers pick up enough ideas, resources and confidence in the topic while they are with their students, they may well decide to have a go back at school in future. We provide loan kits in many subjects as extra resources. We also provide workshops to further equip teachers so that they can run similar programmes themselves. While this may not be good for our visitor numbers, it is surely healthy for education in schools.

The 'Fossils' programme looks set to be a permanent feature of our education menu. Continuing to assess and respond to teacher and student needs will help us to maintain our standing within the education community. One of this institution's strengths is its ability to provide educational opportunities across a wide range of curriculum areas, social studies and art as well as science and technology. Obviously we cannot provide everything that teachers request, but contacts throughout tertiary teaching organisations, research institutes and community groups enable us to be 'knowledge brokers' in our region.

Neville Gardner *is the Senior Education Officer, Science, The Science Centre, Manawatu Museum and Art Gallery, Private Bag 11055, Palmerston North, New Zealand.*

Email: gardnern@pncc.govt.nz

Investing in Education for Better Public Programmes

Brett Dunlop
The Australian Museum, Sydney, New South Wales, Australia

Overview of The Australian Museum

The Australian Museum was established in 1827 and has operated continuously at the College Street, Sydney site since 1845. As a museum of natural history, it has collections, research and public programmes in the areas of cultural heritage and the natural environment. The Museum provides services throughout New South Wales and extends its reach nationally and internationally via its scientific research, web programme and travelling exhibitions.

In the College Street buildings, there are nine semi-permanent exhibitions, two major temporary exhibition spaces, an Education Centre for school groups and a centre for public inquiries and research. The exhibitions and other public spaces in the Museum create environments where visitors of all ages and backgrounds can follow their interests by investigating, exploring, enjoying, questioning, discussing, sharing and learning.

We aim to create a variety of experiences for our visitors. Some experiences will be open-ended; others are structured. Some will be self-directed and others staff-assisted. Do we get it right? Not always! But we can avoid potential pitfalls by investing up front in our research and development processes.

Public Programme Development – Investing in Education

An investment of educational input in public programme development will bring rewards. Consideration of the needs and interests of potential visitors should start early.

At the Australian Museum, the vital process of public programme development starts with the selection and testing of ideas for future exhibitions. A team of managers assesses the exhibition proposals. They can initiate proposals as well as receive them from staff or other museums. Importantly, the team is closely involved in market research to gauge the likely public interest of proposals. This 'front end' evaluation informs our decisions but does not enslave us to 'market forces'.

Once a proposal is commissioned for development, a project team is assembled. Team members are selected for their expertise, experience and ability to contribute great ideas. Our teams are guided to value the expertise of every member. Typically a team has a chairperson and representatives with expertise in design, education, science and marketing. Each contributes in their way to creating successful outcomes for the future visitor.

The concept phase

This is the first and most important phase of the project team's creative work because it is when the ideas are formed. The team translates the project objectives into a conceptual structure, which has a central idea or 'key message'. This early phase is the time to explore all the suggested ideas – from bold to boring – before focusing on how it could be best communicated. What are the most attractive themes, storylines and media for communicating the key message to the target audience? What are the possible interpretive strategies for getting the message across?

The proposal is now ready for another stage of audience testing, formative evaluation. Feedback from potential visitors at this early stage – before a single element is built – will help you know if you are on the right track. Or you may discover that you need to develop a much more exciting presentation!

The value of educational expertise at the early stages of programme development should not be underestimated. The educator can assist in clarifying the communication objectives and synthesising a concise 'key message' for the target audiences. These early decisions will help to focus the team throughout the development process.

The development phase

This is the second major phase of the project, and the longest. Approved concepts are now considered in fine detail. Plans, prototypes, models, sketches, copyright, label text, visitor facilities. In fact, all aspects of the project are now developed and documented ready for production.

Earlier front-end and formative evaluation has informed the team of the interests of the target audience. This helps the team to develop the 'hooks' needed to engage the visitors and keep them interested in exploring further. The educator assists the team at this stage by considering the levels of prior knowledge that visitors may have in the topic. By developing a variety of interpretive strategies, the educator helps to create the overall exhibition experience in which visitors will build their new understandings of the topic. How can you make it work for your visitors? Will you try staff-mediated experiences, live theatre or additional research resources for visitors?

Some of the more specialised forms of educational expertise are called on at this stage of a project's development, e.g. evaluation, storyline development, instructional design and interpretive planning. The Australian Museum has been able to direct some of its resources towards employing specialist evaluators, editors and multimedia developers. Other museums can benefit from our investment. We have published a handbook of language guidelines for museum exhibitions (Ferguson, MacLulich and Ravelli, 1995) and many articles on audience research and learning can be found on our websites:
www.austmus.gov.au/educ/learning.htm;
www.austmus.gov.au/biodiversity/6_3.htm.

The production phase

This is the third major phase of the project, where good planning in the earlier phases is rewarded. In this phase, the components are made, tested and installed. Education materials are produced and visitor programmes are refined.

The later stages of exhibition development are perhaps the 'traditional' time for educational input. Even after the content and presentation is locked-in, an educator can attempt some palliative measures or supplementary programmes to make the exhibition partly effective for people's preferred learning styles, or visitors with specific learning needs. However, an investment in educational expertise 'up front' is more effective and makes for a better outcome for all visitors.

Evaluation in this phase is used to measure the effectiveness of the project and to fix any problems identified by observing visitors using (or misusing) your exhibition or its prototype components.

A School of Thought

A discussion of education in museums would not be complete without addressing the special needs of school audiences. An exhibition has to serve many markets. Frequently, school audiences are not considered as the major market for an exhibition. An educator must find ways to make an exhibition relevant, interesting and accessible for the students and teachers who will visit. Often this is achieved with teachers' notes and student worksheets. Also, a range of workshops, hands-on sessions and teaching activities has proved successful.

Educators can help here too, by advising on the needs and motivations of this audience. We should remember that students are motivated in similar ways to our adult and family audiences. All need time (and help) to orient themselves to the venue and exhibition content. All will use the exhibition according to their interests, abilities and prior knowledge. So what can you do?

- You can help the teachers to prepare their students for the visit. If you provide teachers with specific instructions, they can prepare their students beforehand for what they will be doing. This reduces anxiety and facilitates learning.
- You can allow for the variety of preferred learning styles when designing the exhibition and its educational materials. Build in a range of interpretive strategies and present the 'key message' several times, in different ways.

Student worksheets are more popular with teachers than students, but we do not have to design the worksheet for the teacher. Kids need to explore an exhibition before they focus on the specific content. Most kids will want to socialise as much as they want to attend to their allocated task. It is unwise to ignore the social interactions inherent in a school group. So what can you do?

- You can allow them to explore before each group starts a different task. Groups can focus on a key question or object and study it in some depth. The information gathered in the museum should not be repetition of label text.
- You can ask the children to express and record their feelings, preferences and interests. Ask them to think and choose between options; anything but filling in the missing word on the worksheet.
- You can suggest that synthesising and analysing the gathered information should be done back in the classroom, where the social interactions are more controllable.

Summary

An investment of educational input at the early stages of programme development will reap rewards later. Together with other team members, the educator can formulate the tight definition of key messages and multiple interpretive strategies. The key messages from a reference point for the team – a buoy in the turbulent sea of the development process! Get this bit right. Do your evaluation and market testing early and often. Use visitor feedback to steer away from boring options – try to see from their point of view what is relevant, interesting and enjoyable. Utilise educational expertise to find the right medium for communicating the messages, perhaps doing it in many different ways, to maximise the learning outcomes. Research the interesting things that have already been tried and tested in museum education. Finally, think about how the exhibition will work for visitors once it is open.

Reference

Ferguson, L., MacLulich, C. & Ravelli, L. (1995). *Meanings and messages: language guidelines for museum exhibitions*. Australian Museum, Sydney. Available from the Australian Museum Shop.
+61 2 9320 6150.

Brett Dunlop *is the Manager Visitor Services, Public Programs Division, Australian Museum, 6 College Street, Sydney, New South Wales, 2000, Australia.*

Australian Museum Business Services provides commercial consulting services for all aspects of exhibition development and touring, including education programmes.
+61 2 9320 6311.

How Can Science Centre and Museum Education Programmes be Improved? Teachers' Responses to Pre-visit and Visit Questionnaires

Robin Garnett
The Science Centre, Manawatu Museum and Art Gallery, Palmerston North, New Zealand

Pre-visit and visit questionnaires as important communication tools

The Science Centre, Manawatu Museum and Art Gallery prides itself in tailoring education programmes to the requirements of visiting classes. Pre-visit and visit questionnaires are key tools that we use to communicate with teachers.

The institution has five educators on staff who run education programmes for approximately 20,000 students a year. Although we offer teachers the choice of running programmes themselves or paying more and having programmes run by museum educators, teachers almost always choose the second option. These programmes may be science, technology, heritage, Maori or art programmes. They may be based in exhibitions or they may be independent of exhibitions. All are designed to meet teachers' curriculum needs.

When teachers book programmes for their classes they are sent information about the programmes as well as pre-visit questionnaires to fax back to educators. We find that the pre-visit questionnaires are very important in enabling educators to gear programmes to fit teachers' requirements. in the pre-visit questionnaire. Teachers were asked the purpose of the visit, including expected outcomes, as well as how the visit will fit in with current work being done by the class.

We know that teachers bring students to museums for all kinds of different reasons. For example, a teacher may bring a class to study the technology of interactive exhibits, to compare the effectiveness of different media of instruction, to introduce a new topic or for a treat at the end of term. Educators need to be aware of teachers' reasons for coming in order to maximise the benefit for the visiting students. A mismatch between teachers' and museum educators' expectations can lead to disappointments on both sides so we place a lot of importance on establishing a common understanding before a class arrives.

During their visit, all teachers fill in visit questionnaires that give museum educators feedback about the programme. We ask teachers to hand these questionnaires to our Reception staff before they leave, while the programme is fresh in their minds and before they are buried in other paperwork at school. The questions asked cover the appropriateness of the educational material available and the usefulness of the visit overall in meeting expectations.

What teachers tell us about our education programmes

An analysis of responses from teachers between February and May 2000 showed that 98% were satisfied or very satisfied with the education programme they received. The 2% of teachers who were not satisfied all expressed concerns that the particular exhibition they visited was pitched at too high a level (particularly too high a reading level) for their class.

When asked what they particularly liked about a programme, they wrote the following comments:

Teacher's comment	Number of teacher comments
Programme was interactive and activity-based	162
Good links with school programmes	87
Praise relating to specifics of one programme	87
Good presenters	54
Good information and resources	36
Programme was fun/exciting	21
Good programme organisation	17
Programme was easy to understand	16
Good group work	15
TOTAL	**495**

The features of education programmes that teachers say that they like are all features that our museum educators consciously aim to provide. It is encouraging and affirming for them to find that their efforts are recognised and appreciated. The interactive, activity-based nature of programmes receives much the greatest number of positive comments from teachers.

The same teachers were also asked how the programme could be improved. Not all teachers answered the question but those that did gave the following comments:

Teacher's comment	Number of teachers comments
No improvements needed	153
Changes suggested to specific programmes	55
More time	13
Change programme structure	6
More information beforehand	6
Advertise more/ differently	3
More information after visit	2
Cheaper price	2
More information on website	1
Provide coffee for teachers	1
TOTAL	**242**

The majority of teachers did not make suggestions for improving the programmes. Most of those that did make suggestions related them to specific programmes, for example, 'Give children the names of the planets in Maori as well as English' or 'Include information about the history of the discovery of sound'. These comments from teachers reached educators the day they led the programme and enabled them, if appropriate, to make immediate changes to their delivery.

Overall the experience of educators at The Science Centre, Manawatu Museum and Art Gallery is that pre-visit and visit questionnaires from teachers provide invaluable communication tools that allow educators to align their programmes with teachers' needs and to continually improve their delivery.

Robin Garnett *is the Head of Interactive Programmes, The Science Centre, Manawatu Museum and Art Gallery, Private Bag 11055, Palmerston North, New Zealand.*

Email: garnett@pncc.govt.nz

The Science Centre as Living Laboratory

Paul McCrory
Techniquest/University of Glamorgan, Cardiff, Wales, UK

Melanie Quin
Techniquest, Cardiff, Wales, UK

Introduction

Techniquest, Cardiff, Wales.

On 1st May 1995, after nine years in adapted premises, Techniquest moved to its present waterfront building in Cardiff Bay. This facility, the UK's first purpose-built science discovery centre, includes a 1500 square metre exhibition area, together with a science theatre, a planetarium, a public-access laboratory, a discovery room and a computer-based information centre. Techniquest takes pride in the quality and range of its educational programmes, and the high level of reliability and originality of its exhibits – almost all designed and constructed in-house.

Visitor numbers exceed 220,000 a year. Ticket income and the turnover of the trading subsidiary, Techniquest Enterprises Limited, account for 65% of the operating budget.

In addition to its Cardiff presence, Techniquest operates an active outreach programme, both to schools in South Wales and to general public audiences more widely.

The postgraduate experience at Techniquest

The MSc 'Communicating Science' is unique among UK postgraduate courses in science communication. Run jointly by Techniquest and the University of Glamorgan, students are based at Techniquest and the science centre's facilities are their living laboratory and the focus for wide-ranging practical experiences which form an essential part of the MSc course.

The first related courses established in the UK were the MSc in Science Communication at Imperial College, and the Extramural Diploma in Science Communication at Birkbeck College. The main thrust of both of these is the skills and understanding required to provide a bridge between the public and research scientists through print and broadcast media.

By contrast, students on the Techniquest/University of Glamorgan course develop the awareness, understanding and practical skills to communicate directly with a range of 'publics', face to face, in the informal environment of a science centre or science festival. The course is founded upon the 'reflective practitioner' model of professional practice, and mainly uses experiential learning methods. Students learn to reflect on their experience and theoretical inputs, to articulate these reflections and so gain a deeper understanding of them.

Cost/benefit analysis

From September through to the end of May, students follow eight modules that combine seminars/lectures with practical experience in and around Techniquest and the University campus.

For the first few months, students are arguably a net drain on Techniquest's human resources. The balance shifts dramatically early March: in the run-up to National Science Week and the Easter Holidays, the students themselves become innovators, and develop new tools, techniques and styles of communication. For example, as a year-group they devise, organise, deliver and evaluate an overnight camp-in on the first Friday/Saturday of Science Week. Over the four years the course has run, successive groups have refined the interactive activities offered, and honed the operational practicalities, making this event an annual landmark in the calendar of local youth groups and science clubs!

The dissertation phase of the course (June until the end of August) allows students to develop their personal research interests in the field of science popularisation. The Master's dissertation can involve designing and evaluating an activity within a science centre, or as part of a science festival. Alternatively, it might involve observation and analysis of visitors' learning in an interactive centre. Or the project can be of a more theoretical nature, critically exploring contemporary social and scientific issues in science communication. Twenty-two past students have been awarded the MSc degree, four of them with distinction.

Techniquest – a living laboratory

Paul McCrory graduated with distinction in the class of '99, submitting a dissertation entitled: 'The impact of science centre visits on students – An investigation of the professional judgement of teachers'. McCrory based his analysis on the experience both of teachers bringing school groups to Techniquest, and of those in the Bristol area who had visited the Exploratory and were looking forward to the new facilities at **Explore@Bristol** (Explore opens 6th July, 2000 and replaces the Exploratory as Bristol's *science centre)*.

The abstract of McCrory's project appears below:

The impact of science centre visits on students – an evaluation of the professional judgements of teachers

This research investigates the outcomes of a science centre visit on students in terms of the professional judgements of their teachers. The central aim of the study was to identify different types of impact and to develop models which would encompass these reported outcomes. The secondary aims of the dissertation were concerned with how the teachers tried to integrate the visit into classroom work, and the effect of the visit on older secondary school students.

The survey sample comprised 128 teachers who had previously visited Techniquest in Cardiff or The Exploratory in Bristol. Each of these teachers completed a postal questionnaire, and 15 of them were then interviewed by telephone. A smaller national survey was attempted using notices placed in a teaching journal and on the internet, but this produced a very low response.

Although the environment, the processes and the outcomes involved in a science centre visit form an extremely complex and inter-related 'total experience' for each student, two models of the potential outcomes of a visit were developed from the testimony of the teachers. Like all such models they are only useful to the extent to which they clarify thinking and suggest alternative approaches.

The first model ('platform for formal learning') proposed that the impact of visits can be felt in two main domains – long-term cognitive outcomes (by planting memories which can be used to make connections for future learning); and short-term affective gains (increased motivation and interest in science, which can be harnessed by the teacher back in the classroom).

The second model suggested that to include all the rich diversity of outcomes mentioned by the teachers it was necessary to consider five domains – cognitive, affective, conative, behavioural and social. This model also distinguishes between gains made during the visit and those that are made afterwards. The role of the teacher in encouraging these longer-term outcomes is crucial.

Both models were based on an acceptance that given the nature of the environment of a science centre, the possible outcomes of a visit need to be considered in a different framework than those used to evaluate formal education. The freedom of the informal environment will lead to outcomes that are generally not predetermined and are much more diverse that those of formal education.

For the science centre field as a whole, the gain from this project is considerably greater than one more MSc-qualified professional: the 5-domain model has important implications for exhibit and programme development and evaluation, and also in terms of how teachers can maximise the impact of the visit in the cognitive, affective, conative, behavioural and social domains.

The dissertation's final 'Recommendations' section advocates that exhibition and education programmes be developed and presented in such a way as to encourage particular types of outcome. For example:

- Encouraging affective and conative outcomes can support and even encourage cognitive gains though the use of intrinsic motivation, enjoyment, humour, and variety
- However, concentrating on cognitive outcomes and very structured visits may limit or even negate any affective or conative outcomes ('the worksheet phenomenon')
- The use of trained Explainers can increase the quantity and quality of the connections that children make during a visit without significantly affecting their enjoyment
- Science centres and schools should provide more opportunities to allow students to follow up their interest in science, e.g. website activities, science clubs, science magazines, library books, sources of information about science in the news, practical follow-up activities for students. These are all activities that give opportunities for the affective and conative gains from the visit to combine in producing real behavioural changes and increased involvement with science after the visit
- Science centres should attempt to design exhibits and programmes to encourage co-operative learning and teamwork skills – most students are very keen to engage in group activities.

These findings are by no means restricted to the Cardiff or Bristol experiences. They are equally applicable to the policy makers and developers of exhibits and programmes in modern interactive centres worldwide.

Paul McCrory *is a graduate of the Master of Science programme offered jointly between Tecniquest and the University of Glamorgan.*

Email: paul_mccrory@lineone.net

Melanie Quin *is the Science Communication Director at Techniquest, Stuart Street, Cardiff CF10 6BW, Wales, UK.*

Email: melanie@techniquest.org

Section 3

Using Multimedia in Museums and Science Centres

Interactive computer programs and online adventures have become an important inclusion in exhibitions in science centres, museums and galleries. This section looks at some of the successes and shortfalls of digital technology.

The first article in this section by *Ian Allen* and *Brenton Honeyman*, describes the philosophy behind an award winning CD-ROM developed as part of the Australia on CD program – an initiative of the Australian Federal Government. This program allows the user to experience 'real' science and technology through a digital medium. It was awarded the 1997 ATOM Award for the Best Educational Title.

The next article, by *Brenton Honeyman*, discusses issues associated with the World Wide Web, in particular those related to science centre websites. The World Wide Web offers a myriad of opportunities to reach a large number of people who may not be able to take a 'real' visit to a science centre, but virtual experiences may not be pain-free. The final article in this section discusses some of the problems that may be encountered by museums and galleries particularly where the availability of resources may be limited. In their article, *Simon Yates* and *Sharyn Errington* highlight some problems which may arise in the development and implementation of multimedia programs for exhibitions.

One thing is certain – digital technology is here to stay. Most visitors to science centres, museums and galleries expect to be able to interact with a computer-based exhibit. We hope these articles give some insight into the advantages and limitations of computer technology as a tool for reaching a wider audience.

Ingenious! Edutainment via Interactive Multimedia

Ian Allen
Australian Broadcasting Corporation, Science Online, Sydney, New South Wales, Australia

Brenton Honeyman
Questacon – The National Science and Technology Centre, Canberra, Australian Capital Territory, Australia

"No one knows exactly who 'HQ' is. Some say he's an astronaut who changed his life after seeing the Earth from space. Others say he invented a time machine, and saw how bad the future would be if we didn't take better care of the Earth. That's why he set up the *Ingenious!* organisation, with teams in every part of the world. Most I-teams consist of three members, but in Australia the team is one short. That's why you've been invited to join."

This is the opening scene of a CD-ROM produced by Questacon, Australia's National Science and Technology Centre, the Australian Broadcasting Corporation and Radiant Productions. *Ingenious!* features a roving mobile science team, consisting of the player and two other members of the Australian I-Team.

The CD-ROM contains five interactive detective missions, each with an emphasis on applied science – using systematic thinking to tackle the following Australian science challenges:

- In the Northern Territory, we are trying to save one of our most endangered species, a small wallaby called the Mala. Can we do it?
- Millions of small dead pilchards are being washed up on the beaches of southern Australia. What is the cause? Will it threaten one of our most important fishing industries?
- It is time to develop Australia's entry in the World Solar Car Challenge! Can you design and race a car, powered only by the sun, from Darwin to Adelaide?
- The town of Dwellingup in Western Australia is being threatened by bushfires. Can the *Ingenious!* Team use science to overcome this danger?
- The Parkes Radio Telescope in New South Wales is your base as you search the universe for signs of extra terrestrial intelligence. Will your mission succeed?

Each mission is set within the context of a real world problem. The mysterious 'HQ' sets the brief, but how you get there is up to you and your team. Using audio, and a device called a 'mindmap', your fellow I-Team members argue the merits of particular courses of action. The final decision is yours. You can explore their arguments in more detail by clicking on icons that appear on the mindmap as they make their cases. Whichever action you choose, there will be consequences (sometimes decisions made for good reasons don't work out). Sometimes you will have to retrace your steps. Many of the missions include actual false steps made by the real-life researchers. During each mission, players can collect infocards to help inform the decisions they need to take. As each mission is completed, players can print out certificates to acknowledge their efforts; find out about the real story and the scientists behind the real story; look up a file of useful contacts for following up each mission; and file a report recommending future action.

Ingenious! is optimised for players aged eight to 14, but both older and younger audiences will also find it engaging. The title has been designed to work at several levels. Its extensive use of audio and video makes it accessible to players without developed reading skills, yet its underlying sophistication and deep vein of richer content rewards even adult players.

Creating greater awareness of science

The aim of the project was to create a greater awareness of the process of science and its limits. How do scientists come to know what they say they know? The results of science frequently call for changes in the way society should do things. Often the called-for changes have economic or personal consequences for large numbers people. It should be no surprise that scientific findings often trigger political debate, but too often in such debates each side claims its case is backed by 'science'. But what is science? In a healthy modern democracy the public need to have an understanding of what science is, and what it is not.

The cover of the CD-ROM *Ingenious!*

Ingenious! was designed to give its users greater insight into the strengths and weaknesses of the scientific method. It allows players to participate in the process by which scientists go about their problem solving. How do scientists find ways to test competing explanations? The missions featured in *Ingenious!*

are based upon real and contemporary Australian science problems. Few of the stories end with a clear-cut result. More often, players will have accumulated a body of evidence upon which they are asked to make recommendations for future action.

Promoting science and technology learning outcomes

As an example, the 'Racing on the Sun' challenge involves the player in building a solar car from the wheels up, selecting from a range of different shapes, materials, solar cells, motor system, storage batteries and even paint finish! This has to be accomplished within a budget – it is not possible to select the most expensive option from each range – therefore the player needs to balance price and performance in selecting options.

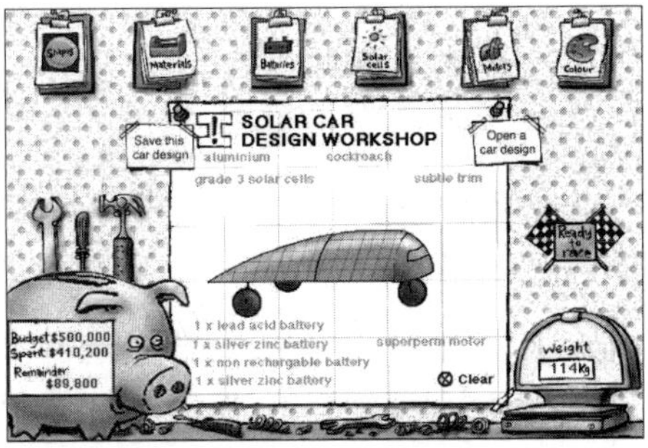

The interactive workshop screen that enables the user to "build" a solar car.

Once the car has been assembled, the player can race it from Darwin in the north of Australia to Adelaide in the south – a distance of 3000 km – using power from the sun. As a full simulation, this challenge can be played again and again as the player experiments with different shapes, materials, solar cells, motor system and storage batteries. In this way, the performance of different designs can be compared in coping with variables such as clouds and hills along the way.

The 'Racing on the Sun' challenge was developed to assist schools in achieving particular outcomes in science and technology curricula.

Optimising educational impact

The project involved a large team of scientists, educators and media specialists. Questacon provided the oversight of the educational content development, working with key organisations such as the Australian Science Teachers Association (ASTA) and the Australian National University's Centre for Public Awareness of Science (CPAS).

As each component of the title was developed, Questacon conducted focus group testing with school children in the target age group as well as with younger children and adults in family settings. This helped to ensure that both the content and functionality of each experience was well suited to its intended audience.

Included with *Ingenious!* is a set of educational resources, enabling teachers to gain direct access to all of the information within each mission, plus comprehensive and classroom-ready notes prepared by the Australian Science Teachers Association. The collaboration with ASTA ensured that this multimedia title would meet the needs of primary and secondary classroom teachers in formal learning environments. Similar collaboration with the Australian National University's Centre for Public Awareness of Science ensured that the *Ingenious!* experience would work well in informal learning environments.

Two copies of *Ingenious!* were distributed free to every school and public library in Australia, as part of the Commonwealth Government's 'Australia on CD' Program. *Ingenious!* was recognised as the winner of the Australian Teachers of Media Award for 'Best Educational Title' as well as the winner of the AIMIA Award for 'Best Entertainment Title' – demonstrating a successful blending of learning and entertainment to create edutainment!

Ian Allen *is the Executive Producer of ABC Science Online, Sydney, Australia.*
Brenton Honeyman *is the Manager, Public Programs and Education, Questacon – The National Science and Technology Centre, King Edward Terrace, Canberra, ACT 2600, Australia.*

Email: brenton.honeyman@questacon.edu.au

Real vs Virtual Visits: Issues for Science Centres

Brenton Honeyman
Questacon – The National Science and Technology Centre, Canberra, Australian Capital Territory, Australia

More and more science centres around the world are issuing an invitation for people of all ages to enter the fascinating world of science and technology through their computer. Science centres are discovering that the Internet – in particular, the World Wide Web – provides a rapidly escalating opportunity to reach a wider audience, thereby promoting a greater public understanding and appreciation of science and technology and their impact on our everyday lives. The capacity of web browsing software to integrate text, images and sounds has attracted many science centres to the idea of establishing a presence on the web. Some science centres are setting out to explore the potential of web-based technologies to deliver experiences, which emphasise interactivity. Indeed, significant developments in web browsing software now make it possible for web users to experience far greater levels of interactivity than the passive, page-turning experiences that dominated the web in the past decade.

Having downloaded a web browser such as Netscape Communicator or Microsoft Internet Explorer, together with a variety of 'plug-ins' which add capabilities to the web browser, the 'visitor' can connect to the world wide web and participate in a wide range of online experiences. Today's web technologies enable the online visitor to interact with sophisticated animated sequences and simulations; navigate in three dimensional environments; control robotic devices on the other side of the planet; view a 360-degree panoramic snapshot of the surface of Mars; and monitor changes in the environment through a 'webcam' which continually captures and updates still or video images.

Several science centres are delivering interactive experiences direct to the school or home via the web. 'Although some of the science centre Web sites lack any useful information beyond museum hours, location and so on, many of them offer enriching experiences, almost like making a real, rather than a virtual, visit.' (May, 1995).

It is this issue of 'real' versus 'virtual' which is causing many science centres to carefully consider the place of online activities and events. The traditional approach of science centres has been to develop visitor experiences based on 'actually being there during opening hours'. In a grid with the dimensions of 'place' and 'time', visitor experiences in science centres would usually be categorised in the 'same place - same time' quadrant at top left (see diagram).

Whenever a visitor interacts with an exhibit, talks to an explainer, listens to a public lecture, watches a science demonstration show or participates in a hands-on activity session, the experience can only take place when the visitor is actually there at the time the experience is scheduled i.e. same place – same time.

Whenever a 'visitor' connects to a science centre web site and interacts with an online activity, reads information, listens to a sound recording, watches a video clip or enters a response as part of an online competition, the experience can take place when the visitor is remote from the science centre and is not restricted by the centre's opening hours or program schedules. These online, virtual experiences would usually be categorised in the 'different place - different time' quadrant at bottom right.

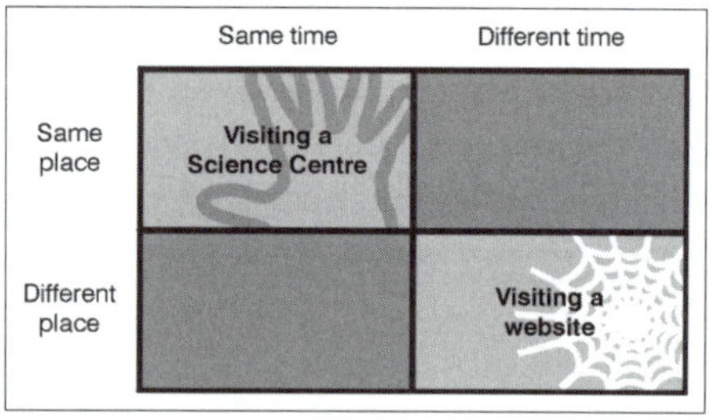

Place-time quadrant

Many areas of everyday activity can also be analysed in terms of the place-time quadrant. For example, the way people have carried out their banking transactions has been to actually visit the bank during opening hours (same place - same time). Banks are now encouraging more and more people to do their banking without actually visiting the bank and at any time of the day or night (different place - different time) through telephone or internet banking services. Whether we like it or not, banks are committed to changing the way we do our banking. By providing experiences for online 'visitors' as well as actual visitors, it can be argued that science centres may reach and influence many more people. This by itself, however, does not provide sufficient reason for establishing an online presence. The purpose of science centres is more than getting a greater number of visitors through their actual or 'virtual' doors.

Providing visitors with first-hand experiences

When the physicist Frank Oppenheimer founded the Exploratorium in San Francisco, he felt concerned that people were becoming information rich and experience poor. He wrote:

"On the whole, people have very little opportunity to have any direct experience with the separate elements of nature or technology. They watch ocean waves, but have never been shown how to observe the way waves pass through each other, bend around corners or bounce off cliffs. In a science museum, one can provide these direct experiences with the behaviour of light, sound and motion. One can set up these experiences in such a way that they not only generate, but partially satisfy curiosity. Science is not just a process of discovering and recording natural phenomena; it is a process which develops our ways of thinking about nature and which enables us to find the connections that simplify and at times enrich our comprehension and awareness of nature." (Oppenheimer, 1968)

Feher (1990), in an article about the role of interactive science centres in studying how people learn, refers to a problem which is endemic in our schools: teachers teach abstractions, definitions and explanations of phenomena that, for the most part, students have never explored, or, worse still, that students may not even know actually occur. If schools so often put the cart (explanations) before the horse (first-hand experience of natural phenomena), modern science museums reverse the process'. Science centres ' . . . present natural phenomena in the form of exhibits that are interactive and manipulable, exhibits whose express purpose is to enable visitors to explore and experiment'. The need of virtual visitors is no different-online activities should enable virtual visitors to explore and interact with phenomena in order to develop first-hand experiences.

Catering for visitor diversity

Actual visitors to science centres bring with them great diversity in terms of their previous knowledge and experience, their assumptions and their expectations, as well as their ways of thinking and learning. Science centres expect this diversity and design their exhibits and programmes accordingly – as open-ended opportunities, which provide flexibility in the manner and level of investigations individuals and groups may wish to undertake. Online exhibits and programs for virtual visitors should also be designed with this principle in mind.

Developing understanding

When actual visitors interact with hands-on exhibits and activities in science centres, they encounter science phenomena in engaging ways, manipulating devices and data to test and develop their understanding – the more compelling the experiences, the more likely learners will develop their understanding of the phenomena.

Virtual visitors, too, are more likely to develop their understanding if online activities are designed to enable people to test their ideas.

What are science centres doing on the web?

A useful place to begin a search for science centres and the various programs they are establishing on the web is a directory of science centres worldwide, located at **www.astc.org**. The Science Learning Network was originally established by six science centres in the USA, each working with a school in their district to develop web-based learning resources and networking facilities for science centre staff, teachers and students. This web site is located at **www.sln.org** and involved the participation of other science centres in other regions of the world.

Like many science centre web sites, the Questacon site provides information about Centre programs and a preview (virtual tour) of Centre galleries. The popularity of this web site is due to Questacon's interest in encouraging online interactivity – after all, Questacon is a centre which encourages as much interaction as possible! Questacon's web site – which features zones for kids, for grown-ups and for teachers – is located at **www.questacon.edu.au**

References

Feher, E. (1990) Interactive museum exhibits as tools for learning: explorations with light International *Journal of Science Education* 1990, Vol 12 (1) 35-49.

Honeyman, B. (1996) *Science Centres and World Wide Web: The Interactive Challenge*. Proceedings of the International Interactive Multimedia Symposium, Perth, 21-25 January 1996.

May, M. (1995) *Scientific surfing for children American Scientist*. November-December 1995, 568-571.

Oppenheimer, F. (1968) *The role of science museums*. In Larabee, E. (ed) Museums and Education Smithsonian Institution Press.

***Brenton Honeyman** is the Public Programs and Education Manager at Questacon - The National Science and Technology Centre, King Edward Terrace, Canberra, ACT 2600, Australia*

Email: brenton.honeyman@questacon.edu.au

Computer-Based Exhibits: A Must-Have or a Liability?

Simon Yates
Graphic and Interactive Software Designer, Canberra, Australian Capital Territory, Australia

Sharyn Errington
Centre for the Public Awareness of Science, Australian National University, Australian Capital Territory, Canberra, Australia

Computer-based exhibits are becoming more common in museums, galleries and science centres and are often expected to be part of exhibitions, especially among the younger generation. The need to include these exhibits can often override the practical aspects of funding, developing and implementing them. Where funding or availability of resources are issues, curators and exhibition designers may need to balance the draw card of having computer-based exhibits against the practical problem of providing a good exhibit.

In mid-1998 a small museum/gallery in Canberra staged an exhibition to show how the community has recycled objects and ideas in and around the Canberra region. The two main aims of the exhibition were to show the many historical and social levels of recycling and to foster community involvement in the museum.

The exhibition consisted of a number of exhibits including:

- national costumes made from recycled materials
- an artistic representation of a slab/bush hut
- computer-generated art combined with cross stitch
- home made toys and objets d'art made from anything at hand
- photographs of the region
- indigenous materials
- photographs of frogs accompanied by digitally enhanced recordings of frog calls
- a multimedia exhibit (*The Armchair Traveller*).

Overall the exhibition was a huge success. This article deals with only one of the exhibits, and problems associated with this exhibit do not reflect on the exhibition as a whole. In this article we discuss the problems involved with the development and implementation of the multimedia program *The Armchair Traveller*. We also provide some advice and guidance to curators of small galleries and museums who may be considering embarking on computer-based exhibits.

The Armchair Traveller: An interactive multimedia program

The purpose of *The Armchair Traveller* was to emphasise the use of combined media and the "recycling" of still imagery into an interactive experience using digital technology. Essentially the program was a user-friendly database of buildings in the Canberra region that have been "recycled" or re-used i.e. whose purpose over the years has changed.

Initially the program was intended as a web site, but subsequently became an intra-net site. The graphic design was pictorial rather than text-based, making it inviting to the user. The program involved a simple "point and click" exercise making it easy to use. The images used were "before and after" shots of various buildings and places that were arranged in a modular structure allowing for future expansion of the program.

The Finished Product

The program worked *in situ* quite well with many more people interacting with it than initially expected. Observations with other computer-based exhibits have indicated that adults tend to be somewhat technophobic and avoid using the exhibits. This was not found to be the case at this exhibition.

For many visitors to the museum, particularly local Canberra residents, the exhibition was directly relevant to their interests. The users felt comfortable with the information and the way in which it was presented. Added to this was the ease of use of the program and its very appealing graphic design.

The Problems

The design and development of *The Armchair Traveller* was fraught with problems, however. Some of these were "in-house" political issues. Others were more general and can be used to provide advice to others who may find themselves in the process of developing exhibition software.

♦ The client–developer relationships

The client's expectations for the finished product were unrealistic in relation to the advantages and constraints of the technology and its delivery, especially in terms of web versus CD-ROM capabilities. It may therefore be necessary to undertake an initial client education process, to ensure an achievable outcome in terms of technology and budget.

- **Know your software**

  To enable development of the program to the client's specifications, the appropriate software used to develop the program was very new. The developers were required to learn the package as well as overcome inbuilt "bugs" in the package. This resulted in the development and design of The Armchair Traveller taking much longer than it should have.

- **Resources: Be prepared**

  In the development of any multimedia program available resources must be considered. In this project there were limited funds available to develop the program to the extent the client had probably hoped. The lack of funds also caused problems in the delivery of the program at the venue.

  Changing perceptions of the client also meant that some re-designs were necessary and the end result was probably not as good as it could have been due to a lack of reserve funding to cater for these changes.

  Originally the computer was situated on a free-standing table in the middle of the exhibition floor which meant there was easy access to computer cables at the rear of the exhibit. This arrangement meant that a member of the gallery staff needed to be in attendance at all times. The aesthetics of the setup were also inadequate for a museum exhibition. For these reasons the computer was later moved against the wall in the venue to avoid any problems with safety and cables becoming dislodged.

  The simple solution to this problem would have been to have the program set up in a kiosk. Lack of funds made this impossible – in fact the computer used to deliver the program was borrowed from the developers for the duration of the exhibition.

  The initial aim during the development of The Armchair Traveller was that it would be expanded beyond the life of the exhibition. Lack of adequate financial support has also prevented this, although other factors such as management problems and lack of technical resources at the museum have played a significant role in the program not being developed further.

Advice to Curators

A sound relationship between the client and the developer is essential for an effective program to be produced that meets the needs and expectations of both the client and the design team. The brief for development must be clear and understood by the developers. If client expectations are unreasonable, this should be communicated to them before development begins.

All members of the development team (which includes the client) must work together coherently to enable the production of a successful program.

Financial constraints are a problem for most museums and galleries and although computer-based exhibits are popular and often expected by visitors, curators should think twice about implementing technology-based exhibits, especially in small galleries/museums where funds are tight.

The development of multimedia programs is expensive. Besides the program itself, appropriate hardware must be available to deliver the program in the best way possible to the visitors – a kiosk is one viable solution.

A program in this situation – with short viewing and interaction times – needs to be easy to use. Avoid the necessity of the audience having to learn how to use the program; it should be intuitive enough to allow a message to be delivered almost immediately.

Technology does fail on occasions and it is imperative to have adequate technical support on hand to deal with any problems that may arise. It can be very frustrating and offputting to visitors to find exhibits out of order so provision for maintenance must be built into the initial budget.

In this project the client and at least one member of the design team were very keen and enthusiastic for *The Armchair Traveller* to be developed. Without this support it would not have happened at all. Despite the problems outlined here, the exhibit did seem to be popular with the visitors. Was it a success, therefore, or not? Are computer-based exhibits a must-have or a liability?

Simon Yates *is the former Art Director of the Multimedia Interactive Learning Laboratory at the Australian National University, Canberra.*
Sharyn Errington *is a Lecturer at the Centre for the Public Awareness of Science Building 42, Australian National University, Canberra, ACT 0200, Australia.*

Email: sharyn.errington@anu.edu.au

Section 4

A Snapshot of Science Centre Research

Research in science centres and museums is increasing all over the world. The articles presented here are snapshots of the kind of work being done and the kinds of conclusions being drawn. All the authors have long research experience and are characterised by work which probes further than simple visitor surveys. We hope that you find them useful.

In the first article, *John Gilbert and Mary Priest* uncover some of the strengths and weaknesses of interactive activity and discussion during a visit of primary students to a science museum. *David Anderson and Keith Lucas* follow the recommendations of Léonie Rennie and Terry McClafferty, made earlier in this book, and develop the theme of effective post-visit activities. *Susan Stocklmayer and John Gilbert* examine adult visitors' responses to interviews about their interactions with exhibits and highlight some of the problems of exhibit design.

All the authors are concerned to describe effective learning in the context of a science-based exhibit. From the point of view of a school based group, the intent of the teacher is usually to promote specific learning outcomes and these can be greatly enhanced by pre- and post-visit activities which are designed carefully. All three articles emphasise the need for engagement by the visitor if learning is to occur. The task for the museum is to provide opportunities for successful engagement and satisfying outcomes.

What do Primary Students Gain From Discussion About Exhibits?

John K. Gilbert and Mary Priest
Department of Science & Technology Education, The University of Reading, UK

Introduction

The number of organised visits by classes, particularly from primary schools (grades 1-7), to museums of all kinds has increased steadily in many countries over the past decade. Museums which emphasise science and technology have been exceptionally busy, perhaps because of the increasingly central place allocated to the subject in the curriculum in many countries. McManus (1992) has distinguished between those science museums organised on a thematic basis, where education is seem primarily as the provision of information (Chambers, 1990) and those organised on a conceptual basis, where the intention is to support the development of specific ideas, in "science centres". While the latter type always involves some physical action by the visitor, there has been a subdivision (Rennie & McClafferty, 1996) into those composed of "hands on" exhibits, which offer no response to the visitor, and "interactive" exhibits, which offer a response and then suggest further activity. It does seem that, in the UK at least, science museum educators are designing galleries that are comprehensive in that they mix the thematic and conceptual approaches and include both hands-on and interactive exhibits.

The wide range of resources held by science museums, relevant to many topics in the curriculum, makes organised visits very attractive to schools. Unique, unusual, and even ordinary objects, replicas, and models are of great potential educational value (Durbin, Morris & Wilkinson, 1990; Gilbert, 1995). The challenge to teachers and museum educators alike is how to realise that potential, and how to improve the quality of learning achieved by pupils. Guidelines for the design, conduct, and follow-up of effective visits to museums, the prerequisites for effective learning, are well established (Bitgood, 1989; Rennie & McClafferty, 1995). While there are many reasons for organising a school visit to a science museum (e.g. improving attitudes toward science, Boyd, 1990), the major issue is how to conceptualise the direct contribution to learning made by such visits. Resnick (1987) differentiates sharply between the nature of "school learning" and "other learning", of which learning in a museum would be an example. Similarly, Wellington (1990) distinguishes "formal learning" which takes place in school, from "informal learning", which takes places in museums. Dierking (1991), on the other hand, argues that such sharp distinctions are inappropriate, seeing the nature of the learning achieved is governed by a number of factors, of which the physical setting is only one.

In recent years, there has been an increased concern to provide "authentic" science education based on activities that as are similar to those experienced by scientists as possible. The characteristics of scientists activities are that they involve inquiry into ill-defined problems by a group of individuals of differing experience working together as a community by a means of a discourse based on the use of shared knowledge and practices (Roth, 1995). Museums can provide contexts that more nearly approximate those of scientific laboratories than do most classrooms. Exhibits typically represent an address to broadly conceived, often implicit, problems from which visitors must construct personal or group concerns. A group of visitors, often composed of individuals of varying experience of the phenomena involved, are able to share prior and present understanding through focused conversation, thus engaging in the social construction of knowledge. Hands-on, and particularly interactive, exhibits provide opportunities for physical activity.

If learning in museums takes place to an appreciable extent through the social construction of knowledge, what is actually involved for those concerned? The study reported in this article views learning as the development and use of mental modes by individuals. Processes involving mental models are shaped by the social context in which they occur. Their production and use are seen to be triggered by events, which, because they have this consequence, are called critical incidents.

Models and Critical Incidents

Models

A model is, in general, a representation of an idea, object, event, process, or system. A model is formed by considering that which is to be represented (the target) analogically in the light of the entities and structures of something which it is thought to be like (the source, Duit & Glynn, 1996). It is used to make predictions, which are then empirically tested, such that the model is subsequently either confirmed, modified, or abandoned. A terminology for the various meanings of "model" has been developed and is used here (MISTRE, 1996).

A mental model is an internal representation of an object, states of affairs, or a sequence of events or processes, of how the world is, and of physiological and everyday social actions. It enables individuals to make predictions and inferences, to make decisions, and to control their execution. By their very nature, mental models can only be directly appreciated by the individuals having them. This means that their nature cannot be directly probed. An expressed model, on the other hand, is that version of a mental model expressed by an individual through action, speech, or writing, and which is therefore available for interrogation. The main characteristic of an expressed model is that it is in the public arena and therefore available for anyone to form a mental model of it.

A consensus model is an expressed model that has been subjected to testing by the academic community associated with a given subject and which has been socially agreed upon by at least some members of the community as having merit for the time being. Thus, consensus models are one of the main products of socially organised scientific activity. One of the main purposes of science education is to have students develop mental models of the major consensus models of science. A teaching model is a specially constructed expressed model used by both teachers and students to aid the understanding of a given consensus model (see, e.g. Treagust, Venville, Harrison, Stocklmayer & Thiele, 1995). Expressed, consensus models are the outcomes of social construction within the academic community, expressed and teaching models are subject to less stringent social verification.

Critical Incidents

An individual is continuously forming, using, and revising mental models. While those models of everyday relevance are activated frequently, those that concern less significant issues will only be used infrequently. Circumstances that provoke such an activation will be noticeable to the individual concerned. They will be distinct incidents in an apparent continuity of experience.

The use of critical incidents in educational research stems from the work of Flanagan (1954) who defined a critical incident in the following way:

"By an incident is meant an observable human activity that is sufficiently complete in itself to permit inferences and predictions to be made about the person performing the act. To be critical, an incident must occur in a situation where the purpose or intent of the act seems fairly clear to the observer and where its consequences are sufficiently definite to leave little doubt concerning its effects" (p.327).

We have taken the view that a critical incident is an event that is suffciently coherent and apparently significant, as reflected in the discourse which takes place, to permit inferences in the form of expressed models, by individuals in a social group.

This definition represents a significant shift from that adopted by Flanagan (1954). In Flanagan's work, what constituted a critical incident, was identified by the individual concerned, whereas here this is done by a researcher. Moreover, the present use is concerned with events that seemed coherent and perhaps significant rather than only those that constituted a major turning point or watershed.

The purpose of the present study is to explore the nature of critical incidents which played an important part in the formation, recall, development, and use of mental models during a well-organised visit to a science and technology museum.

It was decided to conduct the inquiry in a comprehensive gallery of a science museum; that is, one which contained "thematic" elements together with "conceptual" elements (McManus, 1992). This would allow the maximum scope for the deployment of different learning styles by the visiting pupils (McCarthy, 1980). The visit was to be closely integrated into the science curriculum currently being experienced by the pupils in school (Bitgood, 1989). It was considered important for this inquiry that the class teacher involved held the view that good teaching and learning were common in style and form to the school and the museum (Dierking, 1991). By choosing the same personal and physical contexts, in that a class would experience the same curriculum and exhibits, it then would, within the Falk and Dierking (1992) model, be possible to inquire into the significance for learning of the social interactions taking place. The research would, inevitably, focus on expressed models, while seeking to make inferences about mental models (MISTRE, 1996).

Context and study

The study centred on a visit, by a class of 30 pupils aged 8-9 years (grade 4) and their teacher from a state primary school located in a private housing suburb of a small city in the east of England, to the Science Museum in London. The class was just finishing the study of "healthy eating", a topic in the compulsory National Curriculum for Science in England and Wales (Dept of Education, 1995) for that age group. The inquiry was a case study produced by a researcher (M.P) as participant observer, collecting data by interview and observation.

Preparation

The teacher followed her normal sequence of planning for any class visit to a museum. She went to the Science Museum in London and met with the education officers. It was decided to have the pupils visit that section of the Sainsbury *Food for Thought* gallery, which deals with all aspects of the nature, production and marketing of bread. The decision was taken because the school is located in a wheat-growing area, so that the pupils would bring a range of extracurricular experience to the visit, as well as because the gallery contained a good collection of relevant objects and consensus models in material, visual and symbolic forms. The class was not to be required, as is often the case, to complete a worksheet assignment during the visit, but rather allowed to experience the gallery in friendship groups of four each accompanied by an adult. The teacher, a curriculum development officer, museum education officers, an explainer, and the researcher would be available for this purpose. This arrangement would, it was felt, give pupils the maximum opportunity to express and develop their ideas within purposeful groups.

The visit

On arriving at the museum, the pupils were taken to a whole-class activity, staffed by an explainer, which had them examine a grain of wheat closely and explore the properties of flour, particularly the separation of gluten in water. The small groups of pupils, and their accompanying adults, were then free to visit the six chosen stations (exhibits) in the gallery for about one hour in any order that was collectively decided upon within a group at the time; station one (a hand-operated milling machine, or quern); station two (a glass case containing the ingredients of bread and a longitudinal section of a wheat grain); station three (an old fashioned roller mill); station 4 (a set of millstones of varying ages); station 5 (a replica of a modern in-store bakery of the type operated by the Sainsbury Company); and station six (various objects from the history of baking-handling implements, laws about bread, weights and an ancient clay oven).

It had been previously agreed upon that the accompanying adults, while answering any questions asked, would not attempt to instruct the pupils to any significant extent. One of the adults (the researcher) carried a pocket tape recorder to collect as much of the resulting conversation among and with the pupils was possible in a crowded, and inevitably noisy, gallery. The transcriptions of these recordings are used later in this study.

Follow-up

Immediately after returning to school, the teacher made notes of events that could be used as the basis of follow-up activities with the pupils. The teacher's notes were made available for the inquiry. The pupils were interviewed (by M.P.) in an informal way during their science class about their experience of the visit as well as about the follow up activities. These data were transcribed.

Finally, the teacher organised an evening event at the school to which the other pupils at the school, parents and the researchers were invited. Most members of the class had produced a poster about what they had learned during the visit and subsequently. Additionally, a few of the pupils gave verbal accounts of their experiences to the assembled audience, based on their posters. Transcripts of the presentations and the posters were included in the data set.

A full report of the inquiry is available (Priest & Gilbert, 1996). Threads were drawn from the data set so that the particular themes of this study could be addressed.

Critical incidents, discourse, and mental models

The notion of critical incident was used to analyse the discourse about food, and especially about bread, that took place both during and after the visit. Critical incidents were identified by seeking discontinuities, changes of pace and focus of attention, in that part of the overall discourse concerned with bread. This analysis is presented in terms of the discourse initiation, discourse continuation, and discourse closure, instigated by the pupils. Their possible relation to mental modelling was then explored. The examples of discourse quoted used the following convention: T (the teacher), E (the explainer), R (the researcher), and P (a pupil). Pupils are distinguished by number within a quote to show the origin of contributions: there is no continuity of the numbering system between quotes. For convenience of reference, each example of a critical incident is allocated a number. It is not claimed that the categorisation is definitive, but rather that it seems to represent the range of critical incidents occurring with these pupils on this occasion.

Discourse initiation

Three types of incidents were found to be critical in allowing a section of discourse about bread involving the pupils to begin. One example, drawn from a range of examples of each type, is given.

Recognising an object as being familiar

Recognising an object in the gallery as something that had been encountered elsewhere led to the establishment of discourse. For example, two pupils were talking to the researcher about some objects in a display case adjacent to station 5.

CI.1

P1: My Gran's (grandmother's) got some of those weights (pointing at a display case)

R: Does she use them?

P1: No, she uses them as a decoration. My Gran's mother used to use them.

P2: You get some scales – you put those weights on – and the bread – and see how much it weighs – so that you can then sell it.

A discourse about the relation between weight and the sale of bread has been established. This resulted from the recall by one of the pupils of a family situation with which the other pupils readily identified. Mental models seem to have been established about the basis for the division of bread for sale.

Introducing an element of surprise and providing an associated task

The whole-class activity provided by the explainer produced one major discourse, that concerning gluten. The explainer started by throwing handfuls of wheat at the class, which was most unexpected by the pupils. She then told them to examine a wheat grain closely, including its inside structure. This done, she asked the class, as small groups, to mix some flour (which she provided) with water in a bowl. This invitation produced a buzz of interest, which soon settled down in one group into a dialogue involving several people:

CI.2

P1: There's white powder at the bottom – it looks like flour, but there's this grey horrible sticky stuff on my fingers and it won't come off.

E: What about this (holding and stretching the sticky stuff)?

P2: Gluey.

E: It's called gluten and the white powder is called starch (the class continues with the task for some time).

E: Right now, we have divided the flour into starch and gluten. Did you know that gluten makes some people quite ill?

T: Oh, my mum has that and she never eats anything made from ordinary flour. She has to buy special flour with no gluten in it.

P3: What does it taste like, Miss?

T: We'll ask her when we get back. She only lives up the road.

E: The disease has a name. It is called coeliac disease.

This incident was critical both because it allowed new, dramatically different, information to be inserted into the dialogue and because it produced a suggestion for a post-visit activity. The pupils shared a new experience, from which mental models would seem to have been constructed, and were provided with additional information, which evidently consolidated the learning that had taken place. Mental modes about the divisibility of flour into starch and gluten and about limits to the nutritional value of ordinary bread were perhaps involved.

Inserting a question to focus pupils' attention

At station 1, small groups of pupils were allowed to grind wheat in a hand mill (or quern). Initially, they placed all the emphasis on the speed with which they could turn the handle, with scant attention being paid to the production of flour. The explainer caused the emphasis to change:

CI.3

P1: It's really hard work, you've got to do it fast.

P2: It's like an exercise machine.

P3: You have to use the wooden bit sticking up, a handle, really quickly – it's just like an exercise machine for your hands.

E: Why do you use exercise machines?
P4: To build up muscles and make you fit.
E: What do you need to do that?
P4: Energy
E: So you need energy to turn it and make flour.
P4: Oh, I see.

The explainer moved away, leaving the pupils looking at the amount of flour produced at different speeds of mill rotation. A discourse about energy investment in food production was established. Experience had been provided, but, with the help of the explainer's question, mental models seemed to have been constructed that linked these key ideas.

Discourse Continuation

Incidents were critical also for the continuation of a discourse either because they enabled greater meaning to be attributed to past experience or because they acted as a simple bridge to later activity. Some incidents served both purposes. Five types were identified. They were:

- Ideas for Post-visit Activity Are Suggested
- The Generalised and the Particular Are Linked
- Objects Linked Together
- Sustained Attention Provoked
- Text is Successfully Consulted

As an example of the ideas generated for Post-visit Activity, at the end of CI.1 one boy said:

P4: When I get back to school I could try a fair test to see how stretchy gluten really is.

This "fair test" took place, for the word-processed text included in P4's poster at parents' evening said (with some tidying up of spelling to aid intelligibility):

> "We wanted to see how far we could stretch a piece of gluten. We made some dough using flour and water. Then each person took a small piece of dough and tried to wash the starch out of it by squeezing it into a bowl of water. Starch is a carbohydrate and the gluten that is left is a protein. (together) we were left with 700 grams of gluten. Next we tried to stretch the gluten. It stretched to 91cm in 1 minute 3 seconds before breaking."

P4 had been allowed to retain ownership of the idea in that it was he who presented the results to the parents. He had also managed to acquire some additional information, which was included in the text. Mental models, perhaps, initiated at the museum, were extended by providing additional, more focused experience, together with information derived from the relevant consensus model.

Other follow-up activities included: a visit by the teacher's mother to the class; a class visit to a working bakery at the local Sainsbury store: and class practical work arising from the visit.

Discourse closure

Inevitably, most of the bread-related discourse, which started in school before the visit, during the visit, or even in the post-visit activities, came to an unremarked and unobserved close. This would, we suggest, have been because the natural curiosity of the pupils was satisfied or because the teacher decided that the particular section of curriculum had been covered to an appropriate extent. What were much more evident were examples in which discourse came to a premature close, where some event very evidently stopped their continuation. One major type of critical incident seemed to have the following effect.

Unsatisfactory nature of accompanying text

When students approached an exhibit during the visit, their general strategy was to look, perhaps to touch, and only then to look around for information from an accompanying text. Even they often asked their accompanying adult to read it to them while they continued their exploration. As they handled objects while reading took place, the pace of their activity dropped and they seemed to be formulating questions. They expected the text, however read, to answer these questions. In many cases, this strategy evidently worked well: useful answers were acquired and the discourse continued.

However, sometimes the text was not readily accessible to them for technical reasons. It may have been that the gallery was designed with an adult-size audience in mind, for during the post-visit interviews it was said that:

CI.9a

P1: Miss X (the teacher) read some of labels out – she is a bit taller and could see it.

R: Was it hard to see?

P2: Yes, especially for Peter – he's about that big (indicating about half her height).

Some texts were inconveniently positioned:

CI.9b

P1: The writing wasn't near the display enough – Peter was looking at the display and at the nearest thing – it wasn't right – it was too far away from the things.

Some anticipated texts were absent. Thus, at station 4, which was a collection of millstone, the pupils turned to the accompanying diagram to find out more about each of them. In fact, it was a diagram of a waterwheel and, because it had a collection of large stones in the foreground, the pupils mistakenly thought that the display and the diagram were related, to their evident confusion. The situation was made even worse by the fact that, while the diagram had a letter-code labelling system, the key to it was absent. In both cases, additional data, needed if mental modes were to develop further, was not readily available.

While these inconveniences might have slowed the pace of the discourse, there were a few instances in which the text actually interfered with the substance of the discourse. For example, at station 6, a discourse about bread-related laws was proceeding well when the pupils felt the need to consult a panel. This contained both text and a cartoon picture. What happened was as follows:

CI.10

P1: Look, there's something about weights here and laws. (P1 is interrupted by P2, looking at old prints and associated text on the wall)

P2: Oh, look at this funny picture. He's in the stocks and this one is being taken away in a cart (P1 and P2 laugh together and move on).

The data displayed will certainly have met the needs of other visitors. Here, rather than providing information with which the pupils could develop their mental models further, it caused attention to be deflected from those models.

Discussion

The benefits of organising the visit around small groups of pupils, each accompanied by an adult who knew something about bread, were evident in the patterns of discourse that resulted. The adults provided information where this seemed needed (e.g. CI.2); inserted questions that shaped the dialogue (e.g. CI.3); and, most importantly for this group of young pupils, read aloud from text panels (e.g. CI.9). These experts contributed to the putative "community of practice" (Roth, 1985, p.29). The interactions between the pupils themselves played a major role in group learning. Thus, individuals suggested ideas for post-visit activities (e.g. CI.4), acquired information needed in the discourse (e.g. CI.7) and answered questions from their peers about personal knowledge or experience (e.g. CI.6). The social construction of knowledge, mainly mediated by verbal discourse, evidently took place.

Critical incidents seemed to play roles in relation to expressed models and, therefore, by the inferences made above, to mental models. These critical incidents, which initiated discourse, involved the provision of a new activity, focusing an activity onto a meaning which supported the experience being had, or recalling and sharing an established mental model. The raw materials for the mental modelling of experience and the questioning of such models were provided. The critical incidents that facilitated the continuation of discourse were about links. Links were forged between activities in the school and in the museum, between the experiences being had at the museum, between objects on display and consensus models of them and between present experiences and possible future activities. The development of mental ideas were both aggregated and differentiated. The critical incidents that led to the closure of narratives involved an inability to acquire needed information or a deflection from an evident purpose. Perhaps the development of mental models was thereby inhibited.

The conditions provided for the pupils before, during, and after the visit seemed conducive to the production and exploration of expressed models. They had the benefit of the sustained, close, yet non-directional, participation of an adult with some knowledge of the subject being studied (i.e. bread). In Granott's (1993) terms, because there was a considerable degree of symmetry of level of expertise about bread among the members of groups, there was a high level of co-operative learning. Interviews with pupils conducted after the visit, referring to aspects of the visit, coupled with the requirement that pupils produce a poster about the visit for their parents and peers, effectively constituted a stimulated recall of the experiences had. Most important, pupils had multiple and different opportunities to speak, write, and draw about their experience and its significance. The response by pupils to this continuity of experience between school and museum lends weight to the assertion that the types of learning occurring in each are similar (Dierking, 1991).

In recent science education research literature, "teaching models" (or "teaching analogies" as they have otherwise been called), have been portrayed as models specially produced by teachers (or, indeed, pupils), drawing by analogy on objects and events known to the pupils, as an aid to understanding consensus models (see, e.g. Treagust et al., 1995). The present study enables this category to be expanded in two ways. First, many of the wall-mounted texts, which accompanied exhibits, provided a different type of explanation to that evident from the immediate labels attached to objects. Although these young pupils were not able to display "text echo" (McManus, 1989), the reading of these wall panels by the accompanying adults did seem to give pupils access to a different interpretation of events. Second, the authentic objects or replicas on display did seem to act as "exemplary models" in that the more abstract representations of objects, events, or systems (e.g. in the form of diagrams) were exemplified in material form.

The use of a comprehensive gallery, in terms defined earlier, does seem to have been an important ingredient in eliciting on-task discourse during the visit, and hence in promoting the formation, use and development of mental models. Pupils paid various levels of attention to different exhibits, indeed to different parts of any one exhibit. This may have been because different pupils had different learning styles (McCarthy, 1980). To varying degrees, within and between individuals, there may have been preferences to learn by emotional response through concrete experience, by watching through the reflective observation of others, by thinking through the conceptualisation of events, or through doing by experimentation.

A fuller exploration of the social construction of knowledge in museums will entail a closer examination of some of the factors that emerged in this study. It would undoubtedly be insightful to follow one small group of pupils in close detail through their preparation for, experience of, and follow-up to such a visit. Within such a study it would be possible to explore the significance of any discernible mixtures of preferred "learning styles", in McCarthy's (1980) terms, for the choice of exhibit made and for the ensuing discourse. Most interesting of all, it would be instructive to see if the above way of viewing learning in school groups had value for explaining the learning of recreational visitors. Last, a stimulated recall approach, built around the use of video, would enable visitors to directly identify which were the critical incidents for them and, it is hoped, how their thinking changed as a result.

Note: Full details of this study can be found in Science Education (1997) 81(6): 749-62.

References

Bitgood, S. (1989) School field trips: An overview. *Visitor Behaviour*, 4, 3 - 6.

Boyd, W.L. (1990) Museums as centres of learning. *Teacher's College Record*, 94, 761 – 770.

Chambers, M (1990) Beyond "aha": Motivating museum visitors. In *What research says about learning in science museums* (pp. 10-11). Washington, DC: Association of Science – Technology Centers.

Department for Education (1995) *Science in the National Curriculum*. London: Department for Education.

Dierking, L (1991) Learning theories and learning style: An overview. *Journal of Museum Education, 16*, pp. 4-6.

Duit, R. and Glynn, S. (1996) Mental modeling. In G. Welford, J. Osborne and P. Scott (Eds.), *Research in science education in Europe* (pp. 166-176). London, Falmer Press.

Durbin, G., Morris, S. and Wilkinson, S. (1990) *A teacher's guide to learning from objects*. London English Heritage.

Falk, J. and Dierking, L. (1992) *The museum experience*. Washington DC: Whalesback.

Flanagan, J.C. (1954) The critical incident technique. *Psychological Bulletin, 51*, 327 - 358.

Gilbert, J.K. (1995) Learning in museums: Objects, models and text. *Journal of Education in Museums, 16*, 16 - 18.

Granott, N. (1993) Separate minds, joint effort, and weird creatures: Patterns of interaction in the co-construction of knowledge. In R. Wozniak and K. Fischer (Eds.), *Development in context: Acting and thinking in specific environments*. Hillsdale, NJ: Erlbaum.

McCarthy, B. (1980) *The 4MAT system: Teaching and learning styles with right/left mode techniques*. Chicago, Excel.

McManus, P.M. (1989) Oh, yes, they do: How museum visitors read labels and interact with exhibit texts. *Curator, 32*, 174 - 189.

McManus, P.M. (1992) Topics in museums and science education. *Studies in Science Education, 20*, 157 - 182.

MISTRE (1996) *The models in science and technology: Research in education group*. Reading: Department of Science and Technology Education, University of Reading.

Priest, M. and Gilbert, J.K. (1996) *Some contributions to the formal learning of a group of key stage 2 pupils made by a visit to the Sainsbury "Food for Thought: gallery at the Science Museum, London*. Reading,: Department of Science and Technology Education, University of Reading.

Rennie, L. and McClafferty, T. (1995) Using visits to interactive science and technology centres, museums, aquaria and zoos to promote learning in science. *Journal of Science Teacher Education*, 6, 175 - 185.

Rennie, L. and McClafferty, T. (1996) Science centres and science learning. *Studies in Science Education*, *27*, 53 - 98.

Resnick, L.B. (1987) The 1987 presidential address: Learning in and out of school. *Educational researcher*, 13 - 19.

Roth, W. M. (1995) *Authentic school science*. Dordrecht, The Netherlands: Kluwer.

Treagust, D., Venville, G., Harrison, A., Stocklmayer, S. and Thiele, R. (1995) *Teaching analogies in a systematic way*. Perth: Science and Mathematics Education Centre, Curtin University.

Wellington, J. (1990) Formal and informal learning in science: The role of interactive science centres. *Physics Education*, *25*, 247 - 252.

***John Gilbert** and **Mary Priest** are with the Department of Science and Technology Education, The University of Reading, Bulmershe Court, Reading RG6 1HY, UK.*

Email: essgiljo@reading.ac.uk

A Wider Perspective on Museum Learning: Principles for Developing Effective Post-Visit Activities for Enhancing Students' Learning

David Anderson and Keith B. Lucas
Queensland University of Technology, Brisbane, Queensland, Australia

Introduction

Many museum-based studies, which have investigated learning arising from visitors' museum experiences, have restricted such learning to visitors' experiences in the museum itself. Such a view is limiting of actual learning, which has the potential to extend beyond the formulation of knowledge in an in-gallery context. In this paper we discuss a view of learning that recognises the dynamic interplay of visitors' prior knowledge, their experiences during a museum1 visit, and their subsequent life experiences. We discuss some aspects of students' subsequent experiences which teachers are able to capitalise upon, namely planned, classroom-based, post-visit activities.

Defining Learning

A distinction has been made between formal learning, such as might occur in the formal setting of schools, and informal learning, such as might occur in the informal environments of museums and similar locations (Wellington, 1990). However, Dierking (1991) argued that the distinction may not be appropriate because "learning is learning, and it is strongly influenced by setting, social interaction, and individual beliefs, knowledge, and attitudes" (p. 4). In doing so, she stressed the complexity of the learning process in which setting plays an important, but not necessarily dominant role. In similar manner, Hofstein and Rosenfeld (1996) pointed out that informal learning activities, typically developed for out-of-school locations, can be experienced by students in school classrooms. They made the important recommendation that "future research in science education should focus on how to effectively blend informal and formal learning experiences in order to significantly enhance the learning of science" (p. 107).

Schauble, Leinhardt and Martin (1997) contrasted "museum learning" and "school learning" although the distinction is somewhat vague. They appear to equate "museum learning" with learning in museums, but a number of research studies (e.g., Anderson, Lucas, Ginns, & Dierking, in press; Falk & Dierking,

1 We use the term "museum" to include facilities such as history museums, art museums, science centres, aquaria, planetariums, and zoos.

1997) have established that learning may continue well past the time span of a visit to a museum. While acknowledging visitors to a museum may learn from an exhibit, Rennie and McClafferty (1996) posed the question whether learning has occurred if they cannot "link that knowledge to situations beyond their visit" (p. 74). The answer surely depends on the definition of learning that one chooses to adopt. It is clear that researchers interested in learning associated with visits to museums and similar locations need to recognise that learning is multifaceted and unbounded by time, institution or social context.

Explicit definitions of what is meant by the term "learning" have been notably absent from much of the published literature on learning in museums and similar locations during the 90s. For example, Lucas, McManus and Thomas (1986) implied, but did not state, that learning constitutes the acquiring of ideas. If so, research on learning in museums would attempt to assess the amount of information acquired by visitors to museums but Lucas et al. pointed out that "knowing *how* people learn might be more important than knowing *what* they learn" (p. 343, emphasis in original). In some research reports (e.g., Serrell, 1997) it appears that the process of learning is of interest, and in others (e.g., Gilbert & Priest, 1997) the product of learning is the main focus. It is conceivable that the professional background and motivations of researchers play an important role in determining the focus on learning, however, Falk and Dierking (1997, p. 216) have pointed out that learning is neither a process nor a product, but a combination of the two.

In more recent times, references to constructivist views of learning appeared more frequently in the science museum literature (Borun, Massey & Lutter, 1993; Feher, 1990; Hein, 1995; Lucas et al., 1986). A common factor of such references is the recognition of the importance of visitors' prior knowledge, their alternative conceptions, and the individual nature of the construction of meaning from experiences encountered in the museum. The importance of social interactions is recognised to varying degrees. Some researchers have adopted a social construction of knowledge framework for research in science museums (Falk & Dierking 1997; Gilbert & Priest, 1997; Schauble et al., 1997). To do so, they needed to make clear what they meant by "learning." For example, Falk and Dierking (1997, p. 216) asserted that:

> "Learning is the process of applying prior knowledge and experience to new experiences; this effort is normally played out within a physical context and is mediated in the actions of other individuals. In addition, learning always involves some element of emotion and feeling."

This definition highlights the process of learning in the physical, social and personal contexts of the learner. Gilbert and Priest (1997) view learning "as the development and use of mental models by individuals" (p. 751). In this definition of learning, it is the products of learning in the form of mental models, which are highlighted. According to Gilbert and Priest (1997, p. 750), a mental model is "an internal representation of an object, states of affairs, or a sequence of events or processes, of how the world is, and of physiological and everyday social actions." Both definitions are applicable to learning in formal and informal, in-school and out-of-school contexts and, despite the different emphases, may be considered to be complementary and consistent with Ausubel's (Ausubel, Novak & Hanesian, 1978) theory of meaningful learning.

Ausubel's theory of meaningful learning has influenced some research of learning associated with field trips to museums and other out-of-school locations (Anderson, 1999; Balling, Falk & Aronson, 1995; Dierking, 1991; Dierking & Falk, 1994; Falk & Dierking, 1997; Orion, 1993) perhaps because it relates so specifically to the individual's role in making meaning as discrete "bits" of information are added to cognitive structures. Valsiner and Leung (1994, p. 211) describe the knowledge construction process, a synonym for learning, in terms of a transformation of the knowledge structure of an individual. According to Valsiner and Leung, the process is constrained, but not determined by the relationship between the environment and the individual. Their description of the structure of knowledge and transformations between individuals' knowledge states, although framed in Piagetian terms, is highly reminiscent of Ausubel's theory. More recently, Chinn and Brewer (1998) have referred to "snapshots" of what people know at different times, implying the existence of "states of knowledge" and pointed out that researchers "infer that knowledge change is triggered by events involving *new data*, *new conceptions*, *reflection* and *social pressures*." (p. 101, emphasis in original). They claim that "there are few if any comprehensive theories of knowledge acquisition at present. Rather, most current theories are fragments of theories that address one, two or three of the issues" (p. 110).

One theoretical approach that might fill this void is described by its proponents as a "human constructivist" perspective on learning (Mintzes & Wandersee, 1998; Mintzes, Wandersee, & Novak, 1997). The human constructivist view of learning recognises that individuals' present conceptions are products of diverse personal experiences, observations of objects and events, culture, language, and teachers' explanations. Such conceptions are not necessarily consistent with academic knowledge structures. Furthermore, Mintzes et al. (1997) make the important point that "common instructional practices, including those of good teachers and textbooks, are a major source of misunderstanding" (p. 413). This is also true of learning in science museums (Anderson, 1999; Lucas, 1999).

Human constructivism recognises that learning can be at times gradual and assimilative, and at other times rapid and transformative. The former condition implies an incremental change in the individual's conceptual understanding.

The latter condition implies a substantial restructuring of the individual's knowledge. In the words of Mintzes et al. (1997):

> "*conceptual change requires a restructuring of the knowledge framework, and this in turn results from the making and breaking of connections between concepts and sometimes the replacement or substitution of one concept with another*" (p. 415, emphasis in original).

In relation to learning in museums and similar locations, the human constructivist view of learning has potential to guide research and to assist in the interpretation of research data because it recognises the individual's prior knowledge and active personal involvement in knowledge construction, for example during a museum visit. Human constructivism also acknowledges the role played by individuals' present knowledge states, for example as they exit a museum, in determining "the nature and quality of subsequent learning" (Mintzes & Wandersee, 1998, p. 52).

Our view of learning is consistent with human constructivism. We regard learning emergent from an individual's experiences in a museum as a process which is continuous, dynamic and transforming though subsequent experiences beyond the museum setting, and the products of learning to be subject to transformation long after the individual has left the museum.

The Museum Experience

Our understanding of the nature of learning associated with a visit to a museum leads us to reiterate several propositions about such learning.

- A visitor's prior knowledge, culture, attitudes, and beliefs influence significantly how that person will experience and interpret exhibits, events and social interactions within the museum.
- The physical nature of the exhibits and the museum context in which they are presented affect the type and duration of visitors' interactions with exhibits in a museum, and there is a wide range of such interactions by visitors.
- Visitors to museums are frequently members of small groups of people linked by family, social, or educational affiliations. In such circumstances, the nature of visitors' interactions with exhibits, and in particular their interpretation of the exhibits will be mediated in substantial and largely unpredictable fashion by social factors operating within the group.
- New knowledge and understanding resulting from a visit to a museum constitute a foundation for further learning as the individual interprets subsequent events and experiences by reference to the museum experience.

In relation to the last of these propositions, the subsequent events and experiences may derive from sources such as conversations, television programs, movies, reading, travel and, in the case of school students, planned or incidental classroom activities. The potential for extending or enhancing students' learning from visits to museums by planned school-based post visit activities has not been recognised and exploited by teachers (Anderson, 1999; Bitgood, 1989).

We turn now to a discussion of principles for developing such post visit activities (PVAs) that are consistent with our view of learning. The principles outlined were influential in the development of the PVAs used in Anderson's (1999) study of elementary students' learning about electricity and magnetism associated with a visit to an interactive science museum and subsequent post visit activities in the classroom.

Principles for the Development of PVAs

A post-visit activity is a classroom-based activity or exercise specifically designed to enhance learning about a particular topic or phenomenon experienced by students during a visit to a museum or similar informal learning environment. In practice, there are many forms that a PVA may take, and there are many perspectives from which a teacher might develop a PVA (Anderson, 1998). A PVA may be as simple as a classroom-based discussion or as elaborate as an extended research project. Our view is that PVAs have the potential to be highly influential and powerful knowledge building strategies when they recognise the idiosyncratic nature of the learning outcomes for individual students as a result of their museum experiences and enable students to consolidate and enhance their knowledge and understanding in appropriate ways. By "appropriate" we mean consistent with the human constructivist view of learning discussed previously.

Principle 1

Post-visit activities should be built upon students' experiences during their visit to the museum and their pre-existing knowledge, understandings, and related learning experiences so as to consolidate and/or extend their understanding of the themes portrayed in the galleries and their classroom-based curriculum.

It is reasonable to assume that students' understandings of at least some of the information and principles portrayed by the exhibits will be developed in varying degrees as a result of their museum experiences, but the extent of such transformations is difficult for teachers to predict. Nevertheless, the types and extent of knowledge development can be determined in part after museum experiences through a variety of means, such as in-gallery interviews, focus groups, surveys, and similar techniques (Falk & Dierking, 1992; Guba & Lincoln, 1989; Rennie & McClafferty, 1996). With experience, teachers may build a personal awareness of the range and probability of learning outcomes for students whom they have taken to visit a local museum.

Teachers might also gain such insights in a more naturalistic manner during classroom-based debriefing sessions immediately following class visits to the museum. For example, identifying and discussing interesting or puzzling exhibits, or experiences which students found to be memorable is likely to provide the teacher with indications of what students know and understand about the museum exhibits. Such teacher-facilitated actions constitute PVAs which can promote ongoing knowledge construction and reconstruction. In addition, when teachers and museum educators have a sense of what their students actually learn during the museum visit, which frequently is a limited and/or distorted version of the learning outcomes intended by the exhibit designers, they can begin to develop PVAs which capitalise on the students' learning in appropriate ways.

Principle 2

Post-visit activities should be designed in the light of contextual constraints of implementation time, preparation time, availability of resources, and the educational contexts in which students and teachers operate within and beyond the formal school environment.

A class visit to a museum can be considered to be an extension of the formal school experience and is frequently planned accordingly, despite the informal setting and free choice environment of the museum. In like manner, classroom-based post visit activities can be considered to be an extension of the informal museum visit, and the potential benefits of so doing have been identified (Bitgood, 1991; Griffin, 1998; Griffin & Symington, 1997). We argue that, from a teacher's point of view, there are definite benefits for designing PVAs that are closely linked to the formal school curriculum. Linking the experiences to the curriculum provides the advantage of an established relevant context to which the students' experiences in the museum and subsequently in the classroom can be related meaningfully (Anderson, 1998; Bitgood, 1989; Griffin, 1998; Wolins, Jensen, & Ulzheimer, 1992.

In a typical school context, teachers are constrained by time and resources, and perhaps their own knowledge, when seeking to develop and facilitate educationally effective PVAs (Griffin, 1998). However, many activities commonly incorporated in the classroom are suitable for modification as PVAs, it being the appropriateness of the link with students' experiences in the museum rather than the specific nature of the PVA that is crucial. For example, carefully prepared work sheets with fairly specific instructions for students to follow might be quite an appropriate format for PVAs, provided that completion of the work sheets presents opportunities for students to draw upon their personal experiences at the museum in order to enhance their learning, either individually or in concert with others.

Principle 3

Post-visit activities should be related to students' museum experiences and to the broader school-based or other curriculum connected to those museum experiences.

It has been suggested that PVAs should be seen as supporting experiences which help develop students' knowledge and understandings in the light of the wider school, curriculum, and life experiences (Bitgood, 1989, 1991; Griffin, 1998; Javlekar, 1989; Lucas, 2000; Stoneberg, 1981; Wolins et al., 1992). .From a teacher's perspective, PVAs should be developed from the basis of student knowledge which has resulted from the museum experiences, but contextualised within the wider curriculum.

In practice, the third principle requires a teacher to plan a class visit to a museum well in advance of the actual event, and to seek to incorporate as many aspects as possible into the regular curriculum. It also requires the teacher to visit the museum well in advance of the class visit to identify exhibition themes and key exhibits, and to assess their potential for supporting student learning in relation to the broad school curriculum. Part of the process of achieving this is to deconstruct the concepts the designers of the exhibits attempted to convey. Having done so, the teacher will be in a position to organise appropriate pre-visit activities for the students, to anticipate a range of likely student learning outcomes from the museum visit, and to prepare a range of appropriate PVAs.

Principle 4

Post-visit activities should be designed so that they encourage the teacher to respond flexibly to students' emerging and developing understandings and to avoid the PVAs being simply prescriptive in their approach.

Teachers should be sensitive to students' knowledge and understanding so they can direct the PVA in a manner which will provide optimal assistance to students in the construction and reconstruction of their knowledge and understanding. This requires teachers to be both willing and able to adopt a flexible approach to avoid PVAs being simply the completion of mundane tasks set by the teacher. A teacher who is able to respond to a student's knowledge and understanding prior to and during the implementation of the PVA will be likely to provide experiences which are influential in promoting further construction of knowledge and understanding.

Discussion

In this paper we have outlined a view of learning which we regard as being empowering of students' museum experiences beyond the confines of the museum. Further, we presented four principles for the development of PVAs which support and enact this empowerment of students' learning in ways which we regard as being enjoyable, fruitful, and meaningful for students' subsequent learning in school and community contexts. Several practical implications for teachers loom large in the consideration of these principles consistent with the human constructivist view of learning.

Overarching the principles for PVA development is the notion that it vital that PVAs be designed with reference to students' experiences in the museum and to their prior knowledge and understanding. Accordingly, teachers need to attempt to understand their students' knowledge in the domain of the museum exhibition prior to, and emergent during the visit in order to develop effective PVAs.

Because we understand learning to be highly individualistic in nature, we believe that no two visitors will have the same prior knowledge or learning experiences in the museum. Thus it behoves teachers to listen to, and actively seek to explore students' accounts of their responses to the museum and its exhibits. In doing so, teachers will be able to capitalise on the key experiences and unresolved questions of their students in ways that support the wider school-based curriculum. We also believe that is it important for teachers not to be "prescriptive" in their approach to the design and development of PVAs, because what a teacher perceives to be of importance, interest, and relevance may not be congruent with students' interests and needs.

We encourage PVA development that is not only founded on students' prior knowledge and experiences in the museum, but also flexible and divergent in nature. There is immense value in designing PVAs which provide multiple links to a broad spectrum of the school curriculum. In keeping with the human constructivist view of learning, museum visits supported appropriately by such PVAs have the potential to enhance students' learning by providing many opportunities for students to establish links between the museum visit and their school-based experiences, and, indeed to their everyday lives.

References

Anderson, D. (1998, August). *An analysis of the importance of informal and formal science learning contexts to each other: An overview perspective.* Proceedings of the Learning Science in Informal Contexts Conference, Questacon, Canberra, ACT, Australia.

Anderson, D. (1999). *The development of science concepts emergent from science museum and post-visit activity experiences: Students' construction of knowledge*. Unpublished Doctor of Philosophy thesis, Queensland University of Technology, Brisbane, Australia.

Anderson, D., Lucas, K.B., Ginns, I.S., & Dierking, L.D. (in press). Development of knowledge about electricity and magnetism during a visit to a science museum and related post-visit activities. *Science Education*.

Ausubel, D.P., Novak, J.D., & Hanesian, H. (1978). *Educational psychology: A cognitive view*. New York, NY: Holt, Rinehart, and Winston.

Balling, J. D., Falk, J. H., & Aronson, R. (1995, April). *Pre-trip programs: An exploration of their effects on learning from a single-visit field trip to a zoological park*. Paper presented at the annual meeting of the National Association for Research in Science Teaching, San Francisco, CA.

Bitgood, S. (1989). School field trips: An overview. *Visitor Behavior, 4*(2), 3-6.

Bitgood, S. (1991). What do we know about school field trips? *Association of Science and Technology Centers (ASTC) Newsletter, 19*(1), 6-8.

Borun, M., Massey, C., & Lutter, T. (1993). Naïve knowledge and the design of science museum exhibits. *Curator, 36*(3), 209-219.

Chinn, C. A., & Brewer, W.F. (1998). Theories of knowledge acquisition. In B.J. Fraser & K.G. Tobin (Eds.), *International handbook of science education*. (pp. 97-113). Dordrecht, The Netherlands: Kluwer Academic Publishers.

Dierking, L. (1991). Learning theory and learning styles: An overview. *Journal of Museum Education, 16*(1), 4-6.

Dierking, L.D., & Falk, J.H. (1994). Family behavior and learning in informal science settings: A review of the research. *Science Education, 78*, 57-72.

Falk, J.H., & Dierking, L.D. (1992). *The museum experience*. Washington, DC: Whaleback Books.

Falk, J.H., & Dierking, L.D. (1997). School field trips: Assessing their long-term impact. *Curator, 40*(3), 211-218

Feher, E. (1990). Interactive museum exhibits as tools for learning: Exploring with light. *International Journal of Science Education, 12*(1), 35-49.

Gilbert, J., & Priest, M. (1997). Models and discourse: A primary school science class visit to a museum. *Science Education, 81*, 749-762.

Griffin, J.N. (1998). *School-museum integrated learning experiences in science: A learning journey*. Unpublished Doctor of Philosophy thesis, University of Technology, Sydney.

Griffin, J., & Symington, D. (1997). Moving from task-oriented to learning-oriented strategies on school excursions to museums. *Science Education, 81*, 763-779.

Guba, F.G., & Lincoln, Y.S. (1989). *Fourth generation evaluation*. Newbury Park: Sage Publications, Inc.

Hein, G.E. (1995). The constructivist museum. *Journal of Education in Museums, 16*, 21-23.

Hofstein, A., & Rosenfeld, S. (1996). Bridging the gap between formal and informal science learning. *Studies in Science Education, 28*, 87-112.

Javlekar, V.D. (1989). Learning scientific concepts in science centers. In S. Bitgood (Ed.), *Proceedings of the 1989 Visitor Studies Conference* (Vol. 2, pp. 168-179). Jacksonville, AL: Center for Social Design.

Lucas, A.M., McManus, P., & Thomas, G. (1986). Investigating learning from informal sources: Listening to conversations and observing play in science museums. *European Journal of Science Education*, 8(4), 341-352.

Lucas, K.B. (2000). One teacher's agenda for a class visit to an interactive science center. Submitted to *Science Education, 84*, 524-544.

Lucas, K. B. (1999, November). *When Mr Jones took Grade 5 to the Sciencentre.* Paper presented at the annual conference of the Australian Association for Research in Education-New Zealand Association for Research in Education, Melbourne. Retrieved June 12, 2000 from the World Wide Web: http://www.aare.edu.au/index.htm

Mintzes, J.J., & Wandersee, J.H. (1998). Reform and innovation in science teaching: A human constructivist view. In J.J. Mintzes, J.H. Wandersee, J.H, & J.D. Novak (Eds.), *Teaching science for understanding. A human constructivist view.* (pp. 29-58). San Diego: Academic Press.

Mintzes, J.J., Wandersee, J.H., & Novak, J.D. (1997). Meaningful learning in science. The human constructivist perspective. In G.D. Phye (Ed.), *Handbook of academic learning: Construction of knowledge* (pp 405-447). San Diego, CA: Academic Press.

Orion, N. (1993). A model for the development and implementation of field trips as an integral part of the science curriculum. *School Science and Mathematics, 93*(6), 325-331.

Rennie, L.J., & McClafferty, T.P. (1996). Science centres and science learning. *Studies in Science Education, 27*, 53-98.

Schauble, L., Leinhardt, G., & Martin, L. (1997). A framework for organizing a cumulative research agenda in informal learning contexts. *Journal of Museum Education, 22*(2/3), 3-8.

Serrell, B. (1997). Paying attention: The duration and allocation of visitors' time in museum exhibitions. *Curator, 40*(2), 108-125.

Stoneberg, S.A. (1981). *The effects of pre-visit, on-visit, and post-visit zoo activities upon the cognitive achievement and attitudes of sixth grade pupils*. Unpublished doctoral thesis, University of Minnesota, Minnesota, USA.

Valsiner, J., & Leung, M-C. (1994). From intelligence to knowledge construction: A sociogenetic process approach. In R. Sternberg & R. Wagner (Eds.), *Mind in context* (pp. 203-217). Cambridge: Cambridge University Press.

Wellington, J.J. (1990). Formal and informal learning in science: The role of the interactive science centres. *Physics Education, 25*, 247-252.

Wolins, I.S., Jensen, N., & Ulzheimer, R. (1992). Children's memories of museum field trips: A qualitative study. *Journal of Museum Education, 17*(2), 17-27.

***David Anderson** is with the Centre for Applied Studies in Early Childhood*

Email: d.anderson@qut.edu.au

***Keith Lucas** is with the Centre for Mathematics and Science Education*

Email: lucas @pop.qut.edu.au
Queensland University of Technology, Kelvin Grove Campus, Queensland 4059, Australia.

Evaluating the Design of Interactive Exhibits

Susan M. Stocklmayer
Centre for the Public Awareness of Science, The Australian National University, Canberra, Australia

John K. Gilbert
Department of Science and Technology Education, University of Reading, UK

Evaluation in science and technology centres: The background

All science centres have a mission statement, which includes some kind of intent to improve the public appreciation, understanding or awareness of science. Questacon – The National Science and Technology Centre in Australia aspires to be a Centre that "raises national awareness, fosters understanding and instills positive attitudes to science and technology". Aims such as these are, however, impossible to evaluate in a quantitative way, since such evaluation implies exhaustive pre-visit and post-visit testing both of visitors and non-visitors.

Worldwide evaluative studies of science centres have therefore focused mainly on small-scale evaluations of specific aspects of exhibits and exhibitions. Exit surveys (e.g. Groves, 1999), random visitor questionnaires and so on have served to provide some information about affective outcomes but these reports have been used mainly 'in-house', to improve or modify the exhibition itself.

Interactive science centres and museums are, however, increasingly being recognised as centres of informal science learning (e.g. Crane, Nicholson, Chen and Bitgood, 1994; Falk and Dierking, 1992; Falk, Koran and Dierking, 1986; Gilbert, Stocklmayer and Garnett, 1999; Griffin and Symington, 1999; Rennie, 1999; Rennie and McClafferty, 1996; Stevenson, 1991). The influence of "hands-on" activity has been examined in the context of cognitive and affective outcomes through a number of studies (see Griffin and Symington, 1999). These studies have generally sought to describe learning outcomes in terms of the perceived aims of the exhibit, especially in relation to the scientific principles being presented for investigation. The studies have therefore tended to focus largely on more formal outcomes, with the result that visitors have been found to have little recall of the science and, in many cases, little understanding of the principles underlying the exhibit.

A long-term study at Questacon has explored the way in which visitors interact with exhibits in an attempt to explain why engagement with exhibits occurs and what changes occur to the mental models held by visitors (Gilbert, Stocklmayer and Garnett, 1999). To date, with a few distinguished exceptions (e.g.McClafferty and Rennie, 1997; Stevenson, 1991), relatively little psychology-based enquiry has occurred into the nature of the 'interactivity' that takes place, that is into the sensation of learning that visitors experience as they use interactive exhibits. We have chosen to examine this interactivity from the perspective of the visitor rather than that of the science centre.

Turney (1996) suggests that only when people see an opportunity to participate in the conduct of science or, more realistically, to take an active role in learning what science can achieve, will they begin to understand science in the more traditional sense. The Questacon study has emphasised that this active engagement in science (and technology, for the two are intimately entwined in the modern world) is facilitated by interactive exhibits.

The study has established that the learning which occurs is related to the visitor's prior awareness of science and that the exhibit is always interpreted in terms of this awareness. Cognitive and affective outcomes are governed largely by the mental model of the visitor before the interaction commenced, since this essentially dictates the nature and level of the interaction (Gilbert, Stocklmayer and Garnett, 1999, Stocklmayer and Gilbert, in press).

For some exhibits, however, the cognitive and affective outcomes appeared particularly unrelated or only loosely related to the intent of the designer. In these cases, the scientific principles underlying the exhibit usually were presented in a complex analogical form. It seemed likely that the nature of the analogy was contributing to the nature of the outcomes. This has significance for exhibit designers, in that they may be able to predict the likely cognitive outcomes of an interaction in terms of the analogy chosen to represent the underlying science (Gilbert and Stocklmayer, in preparation).

Models in science

An interactive exhibit may be considered as a *model*. Models play a major role in the intellectual and social conduct of science and are substantive outcomes of those activities. A model may be defined in general as a representation of an idea, object, event, system, or process (Gilbert and Stocklmayer, 1997). A model is formed by considering that which is to be represented (the target) analogically in the light of the entities and structures of something which it is thought to be like (the source). The target itself may be termed a 'consensus' model in terms of the following definition (Gilbert, Stocklmayer & Garnett, 1999, p.17):

> "A consensus model is an expressed model which has been subjected to testing by any social grouping, especially by the academic community

associated with a given subject, and which has been socially agreed by at least some members of the group as having merit for the time being."

The underlying ideas – the target – behind an interactive exhibit are most likely to be related strongly to such a consensus model, whether it be a traditional science concept or process or an emerging 'issue' in science.

The exhibit itself, however, is not a consensus model but is a representation of it, designed to facilitate understanding. It is, therefore, a 'teaching' model:

> "A teaching model is a specially constructed expressed model used to aid the understanding of a given consensus model." (Gilbert and Stocklmayer, 1999, p.17).

It is the closeness of the teaching model to the *prior experience* of the visitor which determines the initial engagement, the nature of the experience and the 'success' of the interaction. It is the close or distant *relationship* between teaching and consensus models (the exhibit and the underlying principles) which to some extent determines the overall cognitive outcomes. (Gilbert and Stocklmayer, in preparation; Stocklmayer and Gilbert, in press).

In this research we distinguished between *short-term* outcomes, those achieved during and immediately after the use of an exhibit and *long-term* outcomes, those achieved after sufficient time had passed for the significance of the experience to be considered in a reflective manner. To evaluate long-term outcomes, we interviewed visitors by telephone several weeks after their visit.

Method

The enquiry took place at Questacon – The National Science and Technology Centre in Canberra. Interviews were conducted by a team, which included Questacon "explainers". This aspect of the study was introduced partly to facilitate ownership of the research and to remove the "them and us" idea that often results from external teams of reseachers in science centres, but more importantly to tap into the wealth of expertise which explainers possess and is often unrecognised and seldom reported in research into informal learning.

Initially we called for ten volunteers, of whom five became regular team members. The remaining five fell away through natural attrition, which was expected. The five explainers have become skilled at open-ended interviews after careful inservice and regular feedback about their techniques. The unexpected benefit for them has been an enhanced awareness of visitors' real thinking and, in their view, an ability to carry out their explainer duties more sensitively. It should be emphasised, however, that changing from an 'explaining' style to an unobtrusive interview style, which accepts all answers without the opportunity to 'correct' any perceived misconceptions is very difficult for good explainers. A compromise was reached in that we agreed that, after the interview, there would be an opportunity to discuss the exhibit further and 'help' the visitor where the explainer saw the need. It is accepted that, to some degree, this affects the follow-up interview but, since our intent is to evaluate awareness rather than to evaluate formal learning or exhibit design – and explainers are part of the Questacon experience as a whole- we must accept this modification. It is clearly unreasonable to ask explainers, whose whole ethos is to assist in understanding the concepts underlying exhibits, to allow visitors to move on with what are seen as serious misconceptions or, in some cases, only partial experience of the phenomenon.

A random selection of visitors, roughly representative by age, sex, and social-grouping-at-the-time, was informally interviewed immediately they had completed their use of one (or more) of these interactives. Interviews were audiotaped. We chose to focus mainly on adults or on older children within family groups.

A sample of those interviewed were re-interviewed by telephone some six weeks afterwards, to establish their general views on Questacon and on the exhibits that they had used, and again some three weeks later still, to identify any thoughts and actions that had been prompted by the visit. Some 150 interviews have been conducted during the two phases of the study to date.

We have considered two kinds of exhibit. One kind consisted of popular interactive exhibits which were in the category described by Rennie and McClafferty (1996, p. 56) as "self-contained and decontextualised, with reference to the real-world application peripheral to the exhibit." We have listed examples of these in Tables 1 and 2.

In Table 2, the exhibits make use of analogical representations, which are not closely linked to the target concepts which the designers identify as the desired learning outcomes.

The exhibits in Table 2 require of the visitor that they understand both phenomena – the analogical model and the target concept. This two-step process is facilitated by appropriate graphics, which explain the underlying concept.

The second kind of exhibit in this phase constituted a complete thematic exhibition called *Whodunit?* which, has a murder as the main story line and is essentially a forensic science exhibition. We were interested to see whether the more overt real-world context of a thematic exhibition made a difference to the

Table 1: Types of exhibit construction – direct analogy

Models of reduced or enlarged scale. **"Tornado"**

The Tornado explains the production and movement of a spiral of air moving circularly and transversely at high speed. It represents the phenomenon by a column of water vapour produced by a generator. This column is sucked upwards by an extractor fan at the top. The column is flanked by four vertical tubes through which lateral air jets produce the circular motion. The number of observers and their distribution around the spiral influence its pattern and its effectiveness. Visitors are advised (on the graphic panel) to cluster around in groups.

Models which provide whole-body sensations (a "ride"). **"Earthquake"**

The Earthquake model is a platform on which several visitors sit whilst it is rocked to and fro by a roller mechanism beneath it. The visitor is intended to gain a simulated kinaesthetic experience of what it would be like to be present on ground which is "quaking". The exhibit is usually staffed (for safety's sake) by an explainer, who provides an explanatory commentary tailored to the evident interests of the participating visitors

Models which provide for simple observation of a phenomenon. **"Polarised Light"**

The Polarised Light exhibit invites visitors to place a polariser in front of a beam of light and to observe various objects which include plastic forks, rulers etc. and patterns made by plastic tape. The consensus model of polarisation (wave model of light) is the model underlying this exhibit. The user is invited to discover that different orientations of the polariser cause different patterns in the field of view

Models which provide simple experimental principle apparatus to investigate a **"Roller Race"**

Roller Race relates to the rotational inertia of circular objects when rolling down an incline, depending on the distribution of the mass about the axis of rotation. Three objects of the same mass, respectively a ball, a disc, and a ring, are simultaneously released down an inclined slope. The visitor is invited to predict the order in which they will reach the bottom, to observe and then to explain the result.

way in which exhibits are discussed. *Whodunit* requires the visitor to interact in a number of different ways. The object of the interaction is to test various clues. For example, DNA, tyre treads and fingerprints of suspects can be compared, a cadaver can be examined in layers, and so on. Each exhibit requires concentrated attention to the information and its interpretation and the entire

Table 2: Examples of exhibits which are less closely modelled on the target concept

Example	Desired target concept
Light Harp The Light Harp consists of a series of small holes in which are embedded photo-sensors. About 1m above the holes are a series of light emitters whose beams may be interrupted by the user's hands and arms. As soon as this occurs, musical sounds are heard and the user may "play" the Light Harp by moving the hands across the space above the holes. The sounds are electronically generated and the user has the option to change the nature of the sounds to correspond to various musical instruments, birds cheeping, and so on.	Rectilinear propagation of light/photoelectric effect
Black Hole The Black Hole consists of a metal bowl (representing the gravitational field) into which metal balls (representing any nearby object) are dropped. The balls fall down the sides of the bowl, gaining speed as they do so, and 'vanish' down a hole at the bottom (the Black Hole itself).	The 'Black Hole', an area of space dominated by an apparently annular region of immense gravitational field.

exhibition takes time to traverse. It is less easy to interview visitors after a single exhibit interaction because of the focused nature of the whole experience and some difficulty attached to interrupting trains of thought, so we chose to interview at the exit point.

Results

In summary, we found that the choice of exhibit was indeed often influenced by existing awareness of the topic concerned (Gilbert, Stocklmayer and Garnett, 1999). For example:

I. *What do you think of the Polarised Light exhibit?*
V. *...it was involved in my business, photography and printing, that's what I used to do...But when there was a mention of polaroid for cameras...I was more interested in that aspect of it*
I. *Have you used anything like this before?*
V. *...as a yachtsman I always made sure I had a pair of polaroid sunglasses when I*

was sailing in shallow water...it cuts down the glare and you can see through the water below for any reefs (Male, 60-69 yrs)

There is no doubt that exhibits which model real-world phenomena such as 'Tornado' and 'Earthquake' arouse empathy with those who have already experienced the phenomenon 'for real'. This sense of empathy can relate to personal experience or to distant places:

> *I have friends in the US and they are having a fairly serious tornado season this year... tornados just fascinate me and I had to go and see the movie...*(Male, 40-49 years, a grain farmer).

and it encourages interaction.

Some visitors are drawn to the possibility of a new experience. This applies to Earthquake, for example:

> *I have never experienced an earthquake – I found out that I can stand up to a 5g earthquake like that and shake back and forth* (Male, 60-69 years)

In these situations, information delivered either through the behaviour of the observed phenomenon or from the graphics is seen as interesting and memorable and is able to build on existing knowledge:

> *It is useful, a good illustration of the Richter Scale, how it is measured, the fact that it doesn't just go up, but is more of an exponential. I didn't know that before* (Male, 40-49 years)

The visual impact of an exhibit was also a strong reason for choice. For example, for "Black Hole":

> *I liked the way the steel balls were rolling around that hard surface, they were actually rolling when I walked up to it* (Female, 40-49 years)

Imagination was stirred in respect of the underlying phenomenon:

> *I was thinking: its elliptical. I did a second one with two marbles to see what would happen – they crashed into each other. I was surprised that in the centre they spun really fast, round and round, and stayed up for quite a while* (Male, in his teens).

Even if the visitor to some of the exhibits in this group has little idea of the science, the opportunity for repeated experimentation sometimes proves compelling, as three children aged 9, 12 and 15 (female (f), male (M) and female (F) respectively) explained about Polarised Light:

Interviewer:	*What do you think of it?*
Visitors:	It's cool, it's interesting.
Interviewer:	What's interesting about it? Why is it cool?
F:	You can just stick anything on it, like my mom's glasses, she put them on there and it went all different coloured, psychedelic and stuff.
Interviewer:	So you like it because of these colours?
F:	Yeah, it's just interesting, what it does, what happens with everyday objects.
Interviewer:	So what does happen?
F:	The....I'm not sure
M:	I'm not sure...it just looks really......the light reflects off the bits of glass........and polarises.
Interviewer:	What does polarise mean?
M:	It means the light splits up into separate colours.
Interviewer:	And so what are you actually doing when you are using it?
M:	I think this is polarised glass and you look through polarised glass at other pieces of glass with things on them,.....with other reflective or semi-translucent material and you can see all the colours of the light go through them. I think that's it.
Interviewer:	What about you, would you like to add anything to that?
F:	Not really, I just look at it and go WOW; I don't really try and comprehend...
M:	Psychedelic...
Interviewer:	So you really like this exhibit and when you come to Questacon you try and make a point of using this one?
M:	Yeah, it's cool.
F:	We've done this one about five times already, it's just interesting.
f:	It's like bubbles, when you blow the bubbles they have the little rainbows on them.
Interviewer:	Did you find some of these things were more colourful than others?
M:	Yeah, some had more layers of stuff, different layers of sticky tape when they cross over.

Explanations offered by visitors about unfamiliar exhibits were, however, often qualitative, as would be expected from the nature of the exhibits used. Some explanations were descriptive, together with that most primitive of interpretations, the 'law'. All laws described were erroneous! For Roller Race:

> *Heavier things go quicker, it always happens* (Male, in his teens)

If exhibits demonstrate an unfamiliar phenomenon, visitors can draw few parallels with real world experience or knowledge. Sometimes visitors really struggle to make connections, especially where the science language has real-world applications:

> *I had heard of polarisation…through sunglasses, mainly – and just the fact that there's polarity in so many things in nature. I'm trying to apply that understanding to what I'm seeing here… opposite poles.*(Female, 40-49 years, a water-colourist)

Not surprisingly, in cases such as these, understanding of the science behind the phenomenon was at a beginning level. Complex phenomena, such as the roller race, which demonstrates the effect of different distributions of mass around an axis of rotation using a disc, a ball and a ring, are frequently misinterpreted and, if visitors have prior conceptions or misconceptions, these may well be reinforced unconsciously by the graphic text or by selective observation. Many Roller Race observations were interpreted in terms of the ball being heavier even though the graphics explicitly stated that the "mass" of the objects was the same. Physical testing of the objects conveyed the illlusion of a heavier ball because the pressure on the hand was different. It therefore came as no surprise to these visitors that the ball won the race but their explanations were given in terms of friction, drag and so on.

School visitors were frequently observed filling in worksheets to the effect that the ball reached the end first, but not reading the graphics or, in many cases, conducting the experiment correctly. This experience was clearly of no value, and led us to wonder whether the outcomes at school were judged "successful" in the light of which boxes are ticked. Good preparation would seem to be imperative in these cases.

Many visitors do read the text, and, a small minority without prior experience, do readily accept and recall the explanation:

> *The ball rolls faster because it can use its energy for rolling fast – the energy is used more for speed than rolling. The disc came second because it needed a bit more energy for rolling and the little cylinder came last because it obviously needed most energy to roll and couldn't keep up the speed as well!… It teaches you the way that different objects use the same energy for rolling and some for speed, depending how the mass is divided up in the different objects* (Female, 30-39 years, graphic designer).

The thematic exhibition: *Whodunit?*

It came as a surprise to us that visitors to this exhibition were able to make few comments about individual exhibits. Their experience was holistic and the pursuing of 'clues' dominated their thinking:

I. So there was nothing in particular that you found interesting, but if you had to go away and tell someone about it, is there a part that you are likely to tell them about?

V. No, probably just the interaction, you know, where you have to work it out for yourself...(Male, 40-49 yrs)

Visitors had all heard of forensic science "mainly from watching programs on television" and became wholly engaged in filling in their 'crime file':

> *It requires them to be persistent, not to discount any of the evidence, not to jump to conclusions, to be very methodical...* (Male, 30-39 yrs, science teacher)

This is not to imply, however, that contributions to awareness and to understanding of the underlying science did not occur, but that they were deeply contextualised in the technological aspects of forensic science and to the 'feel' of the exhibition as a whole. Visitors almost never discussed the science concepts themselves, but explained their experience in terms of learning about what forensic scientists actually do:

> *The exhibition was a whole thing, but they've broken it down into bits and you can see what they actually do at each stage...* (Female, 20-29 yrs)

> *I can now explain each of the aspects a little bit better than I could before.* (Female, 30-39 yrs)

The technological applications are the dominant theme here, allowing visitors to access the science concepts covertly.

A drawback to exhibitions of this type can be that visitors spend much longer in front of each exhibit and queues build up at busy times. Visitors can become very frustrated if they have to wait and this detracts from the affective aspects of the exhibition. Where linear progression through a theme is required, considerations of length of engagement, time to read the graphics and so on become vitally important. Interviews of this type can, therefore, assist greatly in exhibition evaluation.

Longer-term outcomes

In the light of the short-term outcomes it is not surprising that the longer- term outcomes were generally confined to the cognitive and affective domains. What was perhaps more surprising was that, once liberated from the researcher's questions about 'what did it mean to you...' visitors were, in the longer term, inclined to give a wider range of response.

Gilbert, Stocklmayer and Garnett (1999) list a number of different ways that visitors explained long-term outcomes of the visit. In addition, it is evident that many discussions occur between family members:

> *We have thought about the moving dinosaurs. My daughter liked them. I can see the exhibit through my daughter's eyes.* (Female, 30-39 years)

Visual memory was mentioned quite frequently:

> *The kids kept mentioning Lightning...they know what it looks like, have a picture in their minds, could see it* (Female, 30-39 yrs)

Those who had scientific knowledge remembered the exhibit in terms of the concept portrayed:

> *I remember light harp... (it) showed that light travels in straight lines...* (Male, 30-39, science teacher).

Reinforcement of school science was valued by parents:

> *We saw the chemistry show, we talked about that and actually this term at school they had a science show visit at school and my son could remember a lot of what he'd seen in relation to that so he found it really good... the children talk about Questacon a lot.*(Female, 30-39 yrs)

There were many wider reflections of the science centre visit reminiscent of those described by Stevenson (1991):

> *(following my visit to Questacon), scientific phenomena are more interesting, less of a mystery, more understandable in terms of basic principles* (Male, 30-39 yrs)

And, to conclude with the ultimate indicator of affective success in the post-modern world:

> *They were so absorbed (in the exhibits) that my son forgot to feed his tamogotchi. We were in the Questacon shop when he realised that his tamagotchi had died. He didn't cry...* (Female, 30-39 yrs)

Discussion

We believe that this approach to the analysis of the experience of visitors to Questacon is able to illuminate some aspects of exhibit design which hinder or facilitate cognitive and affective outcomes. We believe also that it is not particularly useful to assess conventional learning of science concepts during such a visit, but that the wider picture of what visitors are really gaining is much more important.

Many of the interviews revealed ways in which the experience could be improved. They indicted problems with graphics, with exhibit design and with divergence between the designer's intentions and the public's perceptions. The research framework we have adopted here is clearly useful for exhibit evaluation generally.

More specifically, exhibit designers need to consider these aspects of interaction when designing analogical representations of this type. Clearly, expert or even partially knowledgeable visitors are able to integrate the analogy and the target and to appreciate the aesthetic and cognitive aspects of the representation in a critical way. Novices come away with an entirely different kind of learning. Exhibit designers have two choices – to understand that there will be a large group for whom the analogy is remembered but not the target concept, or to re-think how to represent the material. From the perspective of public awareness, there is no doubt that visitors remember the experience with warmth and confidence and that positive attitudes to science are enhanced by the interaction.

In summary, we assert that a visit to a science centre is likely to contribute positively to visitors' awareness of science. It remains for science centres to decide how to make the most of that contribution, whether it be through major exhibitions or through individual exhibits in a thematic gallery. For school visitors it is imperative to allow for time to explore and to accept that results of interactions will be different for each student. If enhanced cognitive outcomes are the purpose of the school visit, very careful planning, preparation of students and design of worksheets is vitally important. Such preparation facilitates engagement (through prior knowledge), successful interaction and, with good exhibit design, connections to the target.

Acknowledgments

We are grateful to the staff of Questacon – the Australian National Science and Technology Centre – for facilitating this study and to Explainers Carla Giuca, Nick Maniatis, and Janey Wood. We are especially grateful to Jennifer Wanless and Marion Williams, who have led the Explainer team, conducted the bulk of the interviews and have contributed important insights into the nature of visitor experiences.

References

Crane, V., Nicholson, H., Chen, M. & Bitgood, S. (1994). Informal Science Learning. *Dedham, MA: Research Communications Ltd.*

Falk, J. H., & Dierking, L. D. (1992). *The museum experience*. Washington, DC: Whalesback Books.

Falk, J. H., Koran, J. J. Jr., & Dierking, L. D. (1986). The things of science: Assessing the learning potential of science museums. *Science Education*, *70*, 503-508.

Gilbert, J.K., & Stocklmayer, S.(1997). *Models and the Public Awareness of Science*. Paper presented at the Annual Conference of the Australasian Science Education Research Association, Adelaide, July.

Gilbert, J., & Stocklmayer, S., (in preparation) Understanding exhibit design: the influence of analogical modelling and remindings.

Gilbert, J.K., Stocklmayer, S.M. & Garnett, R. (1999). Mental modeling in science and technology centres: What are visitors really doing? In: S.Stocklmayer and T.Hardy (Eds). *Proceedings of the international conference on learning science in informal contexts*. Canberra: Questacon, 16-32

Griffin, J., & Symington, D. (1999). Finding evidence of learning in museum settings.? In: S.Stocklmayer and T.Hardy (Eds). *Proceedings of the international conference on learning science in informal contexts*. Canberra: Questacon, 7-15.

Groves, I. (1999) What visitors say they learn a a science centre. In: S.Stocklmayer and T.Hardy (Eds). *Proceedings of the international conference on learning science in informal contexts*. Canberra: Questacon, 146-152.

McClafferty, T., & Rennie, L.(1997). A *triangulation strategy to measure children's learning outcomes from an interactive exhibit*. Paper presented at the Annual Conference of NARST, Chicago, April.

Rennie, L. J. (1999). Learning science in informal contexts: perspectives from research. In: S. Stocklmayer and T. Hardy (Eds.). *Proceedings of the international conference on learning science in informal contexts*. Canberra: Questacon, 1-7.

Rennie, L. J., & McClafferty, T. P. (1996). Science centres and science learning. *Studies in Science Education*, *27*, 53-98.

Stevenson, J. (1991). The long-term impact of interactive exhibits. *International Journal of Science Education*, *13*, 521-531.

Stocklmayer, S.M., & Gilbert, J.K. (in press) New experiences and old knowledge: towards a model for the personal awareness of science and technology. *International Journal of Science Education*.

Turney, J. (1996). Public understanding of Science. *Lancet*, *347*, 1087-1090.

Sue Stocklmayer *is the Director of the Centre for Public Awareness of Science, Building 42, Australian National University, Canberra, ACT, 0200.*

Email: sue.stocklmayer@anu.edu.au

John Gilbert *is with the Department of Science and Technology Education, University of Reading, Bulmershe Court, Reading RG6 1HY, UK.*

Email: essgiljo@reading.ac.uk

Appendix

Science Centres and Science Museums in the Commonwealth

The science centres and museums listed in these Commonwealth countries conduct science and technology programmes in schools and on site. The websites will enable you to see what programmes are available in your region.

Country	Science Centre/Museum	Web Address
Canada	Science World British Colombia, Vancouver, BC	www.scienceworld.bc.ca
	Fraser-Fort George Regional Museum, Prince George, BC	www.museum.princegeorge.com
	Edmonton Space & Science Centre, Edmonton, Alberta	www.edmontonscience.com
	Calgary Science Centre, Calgary, Alberta	www.calgaryscience.ca
	Saskatchewan Science Centre, Regina, Saskatchewan	www.sciencecentre.sk.ca
	Science North, Sudbury, Ontario	sciencenorth.on.ca
	Ontario Science Centre, Toronto, Ontario	www.osc.on.ca
	Discovery Centre, Halifax, Nova Scotia	www.discoverycentre.ns.ca
	National Museum of Science & Technology Corporation Ottawa, Ontario	www.nmst.ca
	Canadian Museum of Nature, Ottowa, Ontario	www.nature.ca
	iSci, Montreal Interactive Science Centre, Montreal, Quebec	www.svpm.ca
	Biodome de Montreal, Montreal, Quebec	ville.Montreal.go.ca/biodome
Cayman Islands	Cayman Islands National Museum	www.museum.ky
England	Science Projects, London	www.science-project.org
	Natural History Museum, London	www.nhm.ac.uk
	at-Bristol	www.at-bristol.org.uk
	Catalyst, Widnes	www.catalyst.org.uk
	Curioxity, Oxford	www.oxtrust.org.uk/educate/curiox/index.htm
	Discovery, Weymouth	www.discoverdiscovery.co.uk/
	Discovery Outpost, Bracknell	www.bracknellforest.gov.uk/council/departments./leisure/recreation/lookout/
	Earth Centre, Doncaster	www.earthcentre.org.uk
	Eureka! The Museum for Children, Halifax	ourworld.compuserve.com/homepages/Eureka_Museum
	Green's Mill and Centre, Nottingham NG2 4QB	www.innotts.co.uk/greensmill/

Country	Science Centre/Museum	Web Address
England *continued*	Guildford Discovery	www.guildford.ac.uk/discover/home.htm
	Herstmonceux Science Centre, Hailsham	www.science-project.org/herst/
	Intech, Winchester	www.hants.gov.uk/leisure/transprt/intech/
	International Centre for Life, Newcastle	www.centreforlife.co.uk/visacfr.htm
	Jodrell Bank Science Centre, Macclesfield	www.highview.co.uk/00/35/003535.htm
	Kaleidoscope, Milford Haven	www.p-net.co.uk/kaleidoscope/
	Magna, Rotherham	www.magna-online.co.uk
	Lowry Centre – Artworks, Manchester	wwwthelowry.org.uk
	Millenium Dome, Greenwich.	www.dome2000.co.uk
	National Space Science Centre, Leicester.	www.nssc.co.uk
	Natural History Centre, Liverpool Museum	www.nmgm.org.uk
	Science Museum, London	www.nmsi.ac.uk
	Scope, Sheffield Hallam University	www.shu.ac.uk/schools/sci/cse/pus/scope.htm
	Search, Gosport	www.hants.gov.uk/museum/seach/index.html
	Ship Shape, Merseyside Maritime Museum	www.merseyworld.com/museums
	Xperiment!, Manchester	www.edes.co.uk/mussci/xperiment.htm
Northern Ireland	Armagh Planetarium	www.armagh-planetarium.co.uk
Scotland	Satrosphere, Aberdeen	www.ifb.co.uk/~ssphere/
	The Big Idea, Ayrshire	www.bigidea.org.uk
	Sellafield Visitor Centre, Cumbria	www.bnfl.co.uk/index1.html
Wales	Techniquest, Cardiff	www.tquest.org.uk
	Centre for Alternative Technology, Machynlleth	www.foe.co.uk/CAT
India	Birla Industrial & Technological Museum, Calcutta Indian National Museum, Calcutta National Science Centre, New Delhi Nehru Science Centre, Mumbai Visvesvaraya Industrial & Technology Museum, Bangalore Central Research and Training Laboratory (CRTL), Calcutta Science City, Calcutta Shrikrishna Science Centre, Patna Regional Science Centre, Lucknow Regional Science Centre, Bhubaneswar	Indian science centres can be found at: www.ncsm.org

Country	Science Centre/Museum	Web Address
India *continued*	Raman Science Centre, Nagpur	Indian science centres can be found at: www.ncsm.org
	Raman Planetarium, Nagpur	
	Regional Science Centre, Guwahati	
	Regional Science Centre, Bhopal	
	Regional Science Centre, Tirupati	
	Regional Science Centre and Planetarium, Calicut	
	District Science Centre, Purulia	
	District Science Centre, Gulbarga	
	District Science Centre, Dharampur	
	District Science Centre. Tirunelveli	
	Bardhaman Science Centre, Bardhaman	
	Dhenkanal Science Centre, Dhenkanal	
	Digha Science Centre & National Science Camp, Digha	
	North Bengal Science Centre, Siliguri	
	Kapilas Science Park, Kapilas	
	Science Activity Corner, Gwalior	
	Science Activity Centre, Sirsa	
	Kurukshetra Panorama & Science Centre, Kurukshetra	
	Goa Science Centre, Panaji	
Malaysia	National Science Centre, Kuala Lumpur	mastic.gov.my/kstas/psn.htm
	Petrosains, Kuala Lumpur	www.petronas.com.my
Singapore	Singapore Science Centre	www.sci-ctr.edu.sg/ssc/ssc.html
Australia	Questacon – The National Science & Technology Centre, Canberra	www.questacon.edu.au
	Scienceworks, Melbourne	www.mov.vic.gov.au/scienceworks
	The Investigator Science & Technology Centre, Adelaide	www.investigator.org.au
	Scitech Discovery Centre, Perth	www.scitch.org.au
	Queensland Sciencentre, Brisbane	www.sciencentre.qld.gov.au
	Wollongong Science Centre	www.uow.edu.au/science_centre
	Discovery Science Centre, Bendigo	www.discovery.as.au
	Newcastle Regional Museum	www.amol.org.au/newcastle
	The Australian Museum, Sydney	www.austmus.gov.au
	The Queensland Museum, Brisbane	www.qmuseum.qld.gov.au
	The Tasmanian Museum, Hobart	www.tmag.tas.gov.au
	Queen Victoria Museum, Launceston	www.qvmag.tased.edu.au
	Museum of Victoria, Melbourne	www.mov.vic.gov.au
	The Western Australian Museum, Perth	museum.wa.gov.au
	The South Australian Museum, Adelaide	www.samuseum.sa.gov.au

Country	Science Centre/Museum	Web Address
Australia *continued*	Great Barrier Reef Aquarium, Townsville	www.aquarium.org.au
	Powerhouse Museum, Sydney	www.phm.gov.au
	Stromlo Exploratory, Canberra	msowww.anu.edu.au/exploratory
Fiji	Fiji Museum	www.fijimuseum.org.fj
New Zealand	NZ Science Centre & Science Alive, Christchurch	www.science-alive.org.au
	Science Centre & Manawatu Museum, Palmerston North	www.pncc.govt.nz/comps/scmm.htm
	National Science – Technology Roadshow, Wellington	www.roadshow.science.org.nz
	Auckland Museum	www.akmused.co.nz
Kenya	National Museums of Kenya	www.museums.or.ke
Namibia	National Museum of Namibia	natmus.cul.na
South Africa	South African Museum, Cape Town	www.museums.org.za/sam
	Albany Museum, Grahamstown	www.ru.ac.za/departments/am
	Giyani Science Centre, Giyani Hartebeesthoek Radio Astronomy Observatory, Gauteng KwaZuzulwazi Science Centre, Durban Mhala Science Centre, Thuamaharhe Museum of Science & Technology Exporatorium-University of Pretoria Geology Education Museum-University of Natal, Durban Physics Dept Science Centre-Uni of the North, Sovenga Unizul Science Centre-University of Zululand, Richards Bay Wetenskatuin-Potchefstroom University Discovery Centre, Cape Town	All remaining South African centres can be found at: www.sasstec.co.za
Swaziland	Swaziland National Museum	www.sntc.org.sz